SpringerBriefs in History of Science and Technology

The *SpringerBriefs in the History of Science and Technology* series addresses, in the broadest sense, the history of man's empirical and theoretical understanding of Nature and Technology, and the processes and people involved in acquiring this understanding. The series provides a forum for shorter works that escape the traditional book model. SpringerBriefs are typically between 50 and 125 pages in length (max. ca. 50.000 words); between the limit of a journal review article and a conventional book.

Authored by science and technology historians and scientists across physics, chemistry, biology, medicine, mathematics, astronomy, technology and related disciplines, the volumes will comprise:

1. Accounts of the development of scientific ideas at any pertinent stage in history: from the earliest observations of Babylonian Astronomers, through the abstract and practical advances of Classical Antiquity, the scientific revolution of the Age of Reason, to the fast-moving progress seen in modern R&D;
2. Biographies, full or partial, of key thinkers and science and technology pioneers;
3. Historical documents such as letters, manuscripts, or reports, together with annotation and analysis;
4. Works addressing social aspects of science and technology history (the role of institutes and societies, the interaction of science and politics, historical and political epistemology);
5. Works in the emerging field of computational history.

The series is aimed at a wide audience of academic scientists and historians, but many of the volumes will also appeal to general readers interested in the evolution of scientific ideas, in the relation between science and technology, and in the role technology shaped our world.

All proposals will be considered.

John Z. Shi

The Voyage from Vortex to Helicity (1517 – 1969)

A Brief History

John Z. Shi
School of Ocean and Civil Engineering
Shanghai Jiao Tong University
Shanghai, China

ISSN 2211-4564 ISSN 2211-4572 (electronic)
SpringerBriefs in History of Science and Technology
ISBN 978-3-032-17852-7 ISBN 978-3-032-17853-4 (eBook)
https://doi.org/10.1007/978-3-032-17853-4

This Springer imprint is published by the registered company Springer Nature Switzerland AG
The registered company address is: Gewerbestrasse 11, 6330 Cham, Switzerland

Dedicated to Lord Kelvin (1824–1907)

The author with the Lord Kelvin Statue in Glasgow on 12 November 2019.

Preface

The concepts and physical phenomena of vortex, vorticity, knot, chirality, and helicity cover a wide spectrum of different disciplines, including natural philosophy, applied mathematics, mathematical physics, turbulent fluid mechanics, topological fluid mechanics, geophysics, and astrophysical plasma. It is true that no one can be really assured of complete knowledge of the five quantities listed above. Nevertheless, this brief contribution attempts for the first time to bridge knowledge of these five quantities, something not previously presented in a single contribution, although they are so beautifully connected. In addition, the concurrent development of the mathematical concepts of curl and divergence seems to be closely related to the mathematical studies of vortex and vorticity. Emphases are placed on some further elaborations made on the historical development of the concepts of vortex, vorticity, knot, chirality, and helicity, with special reference to their early (or possibly original) conceptual, physical and mathematical bases, and their mutual interrelationships. Because each quantity is a difficult subject, there may be statements in the text not fully supported by rigorous arguments. If so, I apologize in advance and welcome all constructive suggestions for improvement.

Shanghai, China
Thanksgiving Day 27 November 2025

John Z. Shi

Acknowledgements

H. Keith Moffatt confirmed that the notion 'vorticity' was first used by Lord Kelvin. Alain Pumir kindly helped the author with an English translation of Lagrange (1781, pages 695, 696, 713, 716, and 717). Catherine Giltrap, Curator and Head of the University Art Collections, Trinity College Dublin, is thanked for sending the author the photo of the sculpture of James MacCullagh. Nigel Buttimore is thanked for providing the author with the detailed notes regarding MacCullagh's contribution to the curl of a vector function. Virginia Mills, Archivist of the Royal Society of London, kindly found the author the relevant information about MacCullagh. David Pritchard sent the author the picture of the Borromean Rings taken in Pollokshaws Burgh Hall, Glasgow, Scotland and a scan of Chapter 4 of Darrigol (2005). Jian-Zhou Zhu is thanked for having drawn the author's attention to chirality during his talk at Shanghai Jiao Tong University on 26 June 2018. The major part of this book was written during Shanghai's COVID-19 lockdowns in 2022. Gert-Jan van Heijst had encouraged the author to continue to work on the early draft until its completion. Other people have read and commented on the various draft versions of this book, including Yang-Wei Liu, Shu-Hua Dou, Jie-Zhi Wu, Gert-Jan van Heijst, David Pritchard, Lord Julian C.R. Hunt, Chris Garrett, Wen-Nan Zou, Les J. Hamilton, Mohamed Gad-el-Hak, Renzo L. Ricca, and John V. Shebalin. David Pritchard and John V. Shebalin are thanked for their suggestions. Les J. Hamilton has kindly corrected one draft and the final revision's English. Renzo L. Ricca has kindly shared his finding of footnote (1) in Moreau (1961, page 2812) with the author. Emily J. Green is thanked for finding the author Hicks's image. Jeremy J. Gray kindly suggested the author to submit this manuscript to Springer for possible publication as a book. Reviewers are thanked for their constructive comments. Lisa Scalone is thanked for her editorial work. Thanks are due to SpringerNature Briefs for the excellence of their work, and their unfailing courtesy. Bethany Downer, Michele Gibney, Anne McLaughlin, Virginia Mills, Chris Pritchard, Kristin Riebe, and Lauren Stark are thanked for giving permission to reprint images.

Competing Interests The author has no competing interests to declare that are relevant to the content of this manuscript.

Contents

Nomenclature

$\boldsymbol{A}$	The magnetic vector potential
$\mathbf{B}$	The field variable
dv	An infinitesimal volume of fluid
dx, dy, dz	The local displacements
e	The ratio of the distance between two particles
$F(\psi)$	Vorticity
$\boldsymbol{H}$	Helicity
H_c	The cross helicity
H_k	The kinetic helicity
H_m	The magnetic helicity
H_P	The parallel helicity
$\boldsymbol{i}, \boldsymbol{j}, \boldsymbol{k}$	Unit vectors in Cartesian coordinates
$\boldsymbol{r}$	The radius vector
u_i	The Eulerian velocity
u_i'	A turbulent velocity
U_i	A mean velocity
V	The volume of a closed system
x, y, z	The Cartesian coordinates
X, Y, Z	The three quantities
α, β, γ	The components of vorticity
α', β', γ'	The components of vorticity
$\alpha'', \beta'', \gamma''$	The components of vorticity
$\alpha''', \beta''', \gamma'''$	The components of vorticity
ϵ_{ijk}	The Levi-Civita symbol
∇	The gradient operator
$\boldsymbol{v}$	The fluid velocity
ρ	The distance of a point from the axis
$\omega', \omega'', \omega'''$	The angular velocities of the fluid
$\bar{\omega}$	Molecular rotation
ξ, η, ζ	The displacements of a particle or the medium
ζ_{ik}	Vorticity

Chapter 1
Introduction

> *...the most diverse phenomena are subject to a small number of fundamental laws which are reproduced in all the acts of nature.*
>
> *Jean Baptiste Joseph Fourier (1768–1830)*

1.1 A Concise Summary of Various Quantities Discussed

To lead the reader to get a sense of direction, a concise summary of various quantities of interest necessary for an understanding of the main text is given below:

- Vortex: in fluid mechanics, vortex (plural: vortexes or vortices) is a mass of whirling fluid or air
- Vorticity: vorticity, which is a pseudovector, is a measure of the local spin of a fluid element near some point
- Knot: originally, knot is a join made by tying together the ends of a piece or pieces of string, rope, etc., but its application to fluid mechanics (e.g. vortex loops) forms part of a new discipline named 'topological fluid mechanics'
- Chirality: chirality essentially means 'mirror-image (left-handed versus right-handed)'. Fluid motions can be chiral. There are two senses of "chirality": (1) as a general synonym for "left- or right-handedness" and (2) specifically as a molecular property associated with certain turbulent fluids.
- Helicity: within the general context of fluid mechanics, helicity is the extent to which corkscrew-like fluid motion occurs. The mathematical definition of helicity is that it is an 'invariant' (constant in time) of the Euler equations of fluid mechanics. An example of helical flow can be found in the wake of windmills or in astrophysical plasmas.
- Curl: refers to a vector operator describing the rotation of a vector field in three-dimensional space.
- Nabla: the name of the mathematical symbol ∇ for the gradient operator.

J. Z. Shi, *The Voyage from Vortex to Helicity (1517 – 1969)*, SpringerBriefs in History of Science and Technology,
https://doi.org/10.1007/978-3-032-17853-4_1

A brief history of development of the various quantities will be presented in the later Chapters. But first, why are they important? How they are related to each other?

Vortex and vorticity are ubiquitous in all kinds of turbulent fluids. They may not be the same physically and mathematically. However, vortex is a more conceptual quantity or a conceptual description of one phenomenon, while vorticity is a physics-based mathematical quantity. Since turbulence can be described by a random field of vorticity, the words, 'vortex', 'eddy', and 'whirl', have the same physical meanings and can be simply called 'vortex'.

Knots, originated from vortex motions, play a central role in fluid motions and can help us better understand turbulent flow in three dimensions. As an example of a flow which has knotted vortex lines, Maucher et al. (2016) theoretically study the excitation knotted vortex lines in matter waves.

Rotational motions in vortex and vorticity stimulate the study of chiral fluids. Chirality is closely related to helicity. As an example of a chiral flow/fluid, Soni et al. (2019) report the creation of a cohesive two-dimensional chiral liquid consisting of millions of spinning colloidal magnets, i.e. a colloidal chiral fluid.

How does chirality relate to vorticity? Vorticity can be induced by chirality. As an example, Yuan (2019) studies vorticity induced by chiral plasmonic fields. Vorticity may have chirality. For example, Palle (2021) studies chirality of the vorticity of the Universe within the Einstein-Cartan cosmology. Chirality, vorticity and magnetic field are mutually interrelated. As an example, the *6th International Conference on Chirality, Vorticity and Magnetic Field in Heavy Ion Collisions* was held in 2021 at Stony Brook University [for details, see: https://indico.bnl.gov].

Helicity is actually a topological measure of the intertwining of vortices in turbulent fluids. How does helicity relate to chirality? In the general theory of fluid dynamo, for instance, chirality plays a key role in that spontaneous magnetic field growth while mean helicity can be the simplest indicator of this chirality.

Throughout the early studies of the five different quantities vortex, vorticity, knot, chirality, and helicity we will find that they are mutually interrelated. The five different quantities, together with the curl and the mathematical symbol ∇, are generally presented in the chronological order of the discovery and development of each. To give the reader a clear sense of key findings, a summary timeline, in table form, of the various advances, is shown in Table 1.1.

The author's interest in these different quantities above is largely stimulated by his own reviews (Shi 2021b, 2024a, b, c). This contribution was initially motivated by the author's invited talk entitled *Vortex Motion and Turbulence: An Introductory Overview*, presented at *the Symposium on The Fundamentals and Applications of Vortex*, which was held on 23–25 July 2021, Soochow University, China, organized by Yao-Hong Qian. In his e-mail to the author dated 28 August 2021, Jie-Zhi Wu raised two questions: "What is the mathematical definition of vorticity in Thomson (1875/1878)? Who first defines the curl of velocity as vorticity?"

Table 1.1 A summary timeline of various quantities and advances in knowledge of them in turbulent fluid studies. Note that page numbers and equation numbers are inclusive in all cited references

Quantities	Advances and references
Vortex	A pictorial study (e.g. Leonardo's drawing *The Deluge* (1517))
	Conceptual/hypothetical studies (e.g. Descartes (1644, e.g. page 107) and Newton (1687, e.g. page 378))
Vorticity	The possible zero-vorticity condition (e.g. d'Alembert (1752, page 100, line 8; page 107, line 11); Euler (1755a/1757a, page 229, Chapter XXVI, the bottom line); and Euler (1755b/1757b, page 290, page 314)
	Early mathematical expressions for the components of the vorticity (μ, ν) (e.g. Lagrange (1760/1762, page 442, line 8)) and (α, β, γ) of Lagrange (1781, page 714, line 4)
	Other early mathematical expressions for the components of the vorticity (e.g. Cauchy (1815/1827, page 42, line 2 from the bottom), Hamilton (1824/1828, page 85, line 8), and MacCullagh (1839/1846, page 21, equation (c); page 22, line 3 from the bottom; page 23, line 3; page 36, lines 2 and 10 from the bottom))
	The Greek letters ω', ω'', ω''' in Stokes (1845, page 290, equation (1)) anticipate the modern symbols for the components of vorticity
	The early notion 'vorticity' and its definition in Thomson (1875/1878)
∇/Nabla	The mathematical symbol ∇ in Hamilton (1831/1837, page 236, equations (D) and (F)) and Hamilton (1840/1843, page 300, line 5)
Curl	The mathematical expression of the curl of a vector function in MacCullagh (1839/1846, page 21, equation (c)), in which the components are of opposite sign to those of curl
	The early notion 'curl' by James Clerk Maxwell in 1870 (Knott 1911, page 143, lines 1–3)
Knot	The early conceptual studies of knot by Maxwell's worble in 1867 (Knott 1911, page 106) and Thomson (1869, page 244)
Chirality	The early study of chirality in Lord Kevin (1884/1904, page 436–467, Lecture XX; pages 602–661, Appendix H and Appendix I)
Helicity	The word 'hélice' (French) appears in Poisson (1811)
	The early study of helicity in the magnetohydrodynamic invariant by Woltjer (1958a, page 489, equation (3); Woltjer 1958b, page 833, equations (1–6))

1.2 Overview and Motivation

Various types of vortex and vortices are ubiquitous in nature, the universe, and engineering fluid flows (Fig. 1.1). It can be believed that the studies of vortex and vortices have been made by some of the best minds in history.

Generations of scientists working in fluid dynamics have recognized the importance of vorticity (Saffman 1981). There is a huge "*Bibliography of vortex dynamics*, 1858–1956" by Meleshko and Aref (2007). There are already many good review papers on the subject, e.g. a 232-page extensive literature review book by Truesdell (1954), a 96-page excellent literature review in Meleshko and Aref (2007), and a 37-page literature review in Meleshko et al. (2012). Other reviews are inclusive in various Chapters, monographs and papers, e.g. Chapter VI. VORTEX MOTION in

Fig. 1.1 Selected examples of vortex and vortices. Top left: a spiral vortex in the cosmos. (Reprinted from NASA, ESA, S. Beckwith (STScI) and the Hubble Heritage Team (STScI/AURA). Public domain). Top right: a phytoplankton vortex in the Baltic Sea. (Reprinted from NASA Earth Observatory/Landsat. Public domain). Bottom: Hurricane Katrina. (Reprinted from NOAA's Office of Response and Restoration. Public domain)

Lamb (1879, pages 146–172), Saffman and Baker (1979), Saffman (1992), Chorin (1994), Lugt (1996), Darrigol (1998), Pullin and Saffman (1998), Chapter 4 VORTICES in Darrigol (2005, pages 145–182), Nitsche (2006), Wu et al. (2006), Moffatt (2010), Aref (2010), and Meleshko (2010). Falconer (2019) gives an interesting account of vortices and atoms in the Maxwellian era. Saffman's (1992) *Vortex Dynamics* has become 'a *sine qua non* of the subject' (Crowdy and Tanveer 2014, page 377, paragraph 2, line 10). Any further work on vortex and vorticity is the better for whatever degree of closeness it attains to the style, clarity, and thoroughness of Truesdell's (1954) *The Kinematics of Vorticity*.

In his classic book *The Kinematics of Vorticity*, Truesdell (1954, page 59, paragraph 2) writes that

> The first treatment of vorticity occurs in the work of d'Alembert (1749) and Euler (1752, 1755a,b); Lagrange (1760) and Cauchy (1815) were the first to introduce single letters to stand for the vorticity components. All this early work is purely formal and somewhat mystifying. The kinematical significance of the vorticity did not begin to be recognized until MacCullagh (1839) and Cauchy (1841) proved that the components of the curl satisfy the vectorial law of transformation.

Both '*purely formal*' and '*somewhat mystifying*' motivate the author to think that some further detailed studies can be done of the early work such as d'Alembert (1749), Euler (1750/1752), Lagrange (1760), Cauchy (1815/1827), MacCullagh (1839/1846), and Cauchy (1841) in order to better elucidate and describe their conceptual and mathematical developments in vorticity, especially how the mathematical expressions for vorticity and the components of the curl were developed. According to Fraser (1985, page 190, lines 3–8), Lagrange published two papers in 1760 [1760a, b]. In Truesdell (1954, page 59, paragraph 2), Lagrange (1760) actually refers to Lagrange (1760/1762), i.e. 1760 refers to the date of his paper completion/reading and 1762 to the date of his paper publication. Even though Lagrange (1760) and Cauchy (1815) are probably among the first to introduce single letters to stand for the vorticity components, details regarding how they developed their ideas might be further elaborated. MacCullagh's (1839/1846) work is not on fluid motion but on the propagation of light in a biaxial crystal. It would be interesting to delineate how the physics and mathematics in MacCullagh's (1839/1846) work on the propagation of light in a biaxial crystal can be potentially related to vortex and vorticity in turbulent fluids.

Truesdell (1954, page 58, footnote 1) suggests that the name vorticity was introduced by Lamb (1916, 2, preface and §30). Is this really true? If not, who first coined the word 'vorticity'? The mathematical expressions for the components of the vorticity were generally thought to be due to Helmholtz (1867, Tait's English version of Helmholtz (1858, page 490, equation (2)). Is it really true? If yes, how did Helmholtz develop it? If not, who proposed the early mathematical expressions for the components of the vorticity? How were they developed? After looking back and reading those literatures available, e.g. Lamb (1879), Truesdell (1954), Saffman (1992), Darrigol (2005), Meleshko and Aref (2007), and Meleshko et al. (2012), the author has still found some missing points and even details regarding the early physical ideas and mathematical expressions for the components of the vorticity and the notions of vortex and vorticity.

From a historical perspective, are there any relevant works on vortex and vorticity which to some greater or lesser extent inspire the works done by d'Alembert and Euler? Chapter 4 VORTICIES in Darrigol (2005, pages 145–182) mainly focuses on Helmholtz (1858), but Darrigol does make some useful comments about how Helmholtz's work related to earlier ideas, and also why vorticity was neglected for so long.

1.3 Primary and Secondary Objectives

In the beginning, the author attempts to revisit those questions above, further the details, and fill in some missing points and gaps existing in those literatures, e.g. Lamb (1879), Truesdell (1954), Saffman (1992), Darrigol (2005), Meleshko and Aref (2007), and Meleshko et al. (2012). To this end, based on the literatures available, as detailed a historical review as possible is made. Some overlapping between this book and those literatures (e.g. Truesdell 1954; Darrigol 2005; and Meleshko and Aref 2007) is inevitable but will be minimized as much as possible.

Although studies of knots are briefly reviewed in Meleshko and Aref (2007), the early studies of knots and how they are related to vortex and vorticity need to be clarified. Since curl, chirality, and helicity are associated with the development of vortex and vorticity, early studies of them are also briefly reviewed in this book. Their relations with turbulence are also briefly reviewed or discussed. Note that examples of knots, chirality, and helicity will be shown in Chaps. 8, 9, and 10, respectively.

The secondary objective of this book is to situate technical discussions within broader historical, cultural, or scientific contexts, showing how these concepts were embedded in their time. It also attempts to write with historians of science and technology in mind, making explicit why vortices, vorticity, chirality, and helicity are historically meaningful, even for readers outside the specialist community.

It should be pointed out that the early literatures referred to include those not just in studies of fluid motion but also in optical phenomena, in electricity, in magnetism, and in galvanic currents.

1.4 Overall Structures

The overall structures of the remainder of this book are as follows:

- Chapter 2 gives a short account of pictorial and hypothetical studies of vortex and vortices from 1517 to 1687.
- Chapter 3 presents an overview of the early mathematical studies of the zero-vorticity condition of the fluid motion from 1752 to 1757 and also provides a detailed account of the mathematical studies of vorticity of the fluid motion from 1760 to 1827.
- Chapter 4 mainly outlines the mathematical studies of the rays and the propagation of light from 1824 to 1846.

- Chapter 5 highlights the mathematical studies of fluid motion, the magnetic and galvanic forces, and vortex-motions in from 1845 to 1858.
- Chapter 6 briefly reviews the mathematical studies of molecular vortices, vortex atoms, spherical vortex, spiral vortex and notion 'vorticity' from 1851 to 1954.
- Chapter 7 briefly reviews the development of the symbol ∇ from 1831 to 1871.
- Chapter 8 briefly reviews the conceptual and mathematical studies of worble and knot from 1867 to 1885.
- Chapter 9 presents an overview of the conceptual, physical and mathematical studies of chirality from 1856 to 1884/1904.
- Chapter 10 gives a short account of the early mathematical studies of hélice (French), helicity and Schraubensinn (German) from 1811 to 1969.
- Epilogue presents some general conclusions.

Chapter 2
The Early Studies of *Vortex* from 1517 to 1687

All my life through, the new sights of Nature made me rejoice like a child.

Marie Curie, née Sklodowska (1867–1934)

2.1 Pictorial Studies of a Large Scale Spiral Vortex by Leonardo da Vinci in ca. 1517

Leonardo da Vinci (1452–1519) was an Italian polymath, having been a scientist, mathematician, engineer, inventor, anatomist, painter, sculptor, architect, botanist, musician and writer (left in Fig. 2.1).

Leonardo's pictorial studies of turbulence are known long since (e.g. Taylor 1970; Colagrossi et al. 2021; Marusic and Broomhall 2021). Leonardo's pictorial studies of vortices have been briefly reviewed in Section 7. VORTICES AND THE HEART (Marusic and Broomhall 2021, pages 11–12).

As shown in Fig. 2.2, *The Deluge* is one of Leonardo's famous paintings in 1517. Interpretation of turbulence with helicity in this painting is made by Moffatt and Dormy (2019, page 19, Figure 1.4). Overall, Leonardo's drawing *The Deluge* may be regarded as a large scale spiral vortex. The same interpretation of spirals is made by Marusic and Broomhall (2021, page 10, Figure 5; page 14, Figure 9c).

Can Leonardo's drawing *The Deluge* be interpreted as a large 'vortex' or a number of vortices? Since both turbulence and helicity are related to vortex and vorticity, Leonardo's drawing *The Deluge* may illustrate the pictorial conception of vortex as well and can be cautiously interpreted as *a large scale spiral vortex*.

It already emerges from Marusic and Broomhall (2021) that Leonardo dedicated a lot of time to depict turbulent fluid motion using spirals. Nevertheless, for this particular drawing, *The Deluge*, its alternative interpretation as "*a large scale spiral vortex*" should be acceptable and may also be of interest to applied mathematicians (or geophysical and astrophysical fluid dynamicists). The two interpretations, 'turbulence with helicity' and 'a large scale spiral vortex', are conceptually similar or

J. Z. Shi, *The Voyage from Vortex to Helicity (1517 – 1969)*, SpringerBriefs in History of Science and Technology,
https://doi.org/10.1007/978-3-032-17853-4_2

Fig. 2.1 Left: Leonardo da Vinci (1452–1519). (Photo by Nico Barbatelli. Reprinted from Wikimedia (https://commons.wikimedia.org/wiki/File:Leonardo_da_Vinci_LUCAN_Hohenstatt_1_portrait.jpg. Accessed on the 27th of November 2025). Public domain). Center: Renati Des-Cartes or René Descartes (1596–1650). (Reprinted from Wikipedia (https://en.wikipedia.org/wiki/Ren%C3%A9_Descartes#/media/File:Portrait_of_Ren%C3%A9_Descartes,_bust,_three-quarter_facing_left_in_an_oval_border,_(white_background_removed).png. Accessed on the 27th of November 2025). Public domain). Right: the bronze bust of Isaac Newton (1642–1727), Isaac Newton Institute for Mathematical Sciences, the University of Cambridge. (Photo taken by the author)

Fig. 2.2 Leonardo's drawing *The Deluge* (1517). (Reprinted with permission from The Royal Collection Trust (RCIN 912380). All rights reserved)

the same. Within the general context of this review, the latter expression may be better than the former.

2.2 Hypothetical Studies of Fluid Motion by René Descartes in 1644

René Descartes (1596–1650) was a French mathematician and philosopher (center in Fig. 2.1). However, Descartes lived in the Dutch Republic from 1628 to 1649 (Jorink and Maas 2012, page 17, paragraph 2, lines 1–2). During his stay in the Dutch Republic, Descartes published his famous book *Principia Philosophiae* (*Principles of Philosophy*) in AMSTELODAMI (Amsterdam) by the Dutch publisher LUDOVICUM ELZEVIRIUM in 1644. LUDOVICUM ELZEVIRIUM (Latin) refers to Louis Elzevir. The modern publisher Elsevier took its name from Elzevir but is not otherwise connected with the family (it was founded in 1880).

As written in *Principia Philosophiae*, Descartes's study of nature and the universe was made with strong religious connotations. Studies of fluid motion can be found in Descartes (1644, Part Second, pages 62–68, Chapters LIV, LVI, LVII, LVIII, LIX, LX, LXI, and LXII). Descartes studies fluid motion within the general context of *The Principles of Material Things*. In his own definition, '*fluids are bodies whose numerous tiny particles are agitated, moving in all directions*,' (Descartes 1644, page 62, Chapter LIV, lines 2–3 from the bottom). What is Descartes's motive for studying fluids and their motions? In his own words, '*The heavens are fluid*' (Descartes 1644, page 79, Chapter XXIV). Note that those English translations are based on Jonathan Bennett at https://earlymoderntexts.com.

As shown in Fig. 2.3, a sketch showing '*turbulent/eddy-like*' fluid motion can be found in Descartes (1644, pages 63 and 64). However, if we take a close look at this portion, we shall be able to see two dotted ellipses within which a cube-shaped '**B**' is located at the centre.

2.3 Hypothetical Studies of *Vortex* by René Descartes in 1644

Within the general context of *Part Three The Visible Universe*, René Descartes studies in detail '*vortex*' (Latin) (its genitive: '*vorticis*') in five Chapters (Descartes 1644, Part Three, pages 105–109): Chapter LXV The poles of each celestial vortex touch the parts of other vortices which are remote from their poles; Chapter LXVI There must be some deflection in the motion of the vortices so that they can move in harmony; Chapter LXVII Two vortices cannot touch at their poles; Chapter LXVIII The vortices are of unequal size; Chapter LXVIX The matter of the first element flows from the poles of each vortex towards its centre, and from the centre towards the other parts. These English translations are based on Jonathan Bennett at https://earlymoderntexts.com.

Fig. 2.3 A sketch showing '*turbulent/eddy-like*' fluid motion in Descartes (1644, page 64). (Reprinted from the Lessing J. Rosenwald Collection (Library of Congress). Public domain)

As shown in Fig. 2.4 (left), the poles of each celestial *vortex* touch the parts of other *vortices* which are remote from their poles (Descartes 1644, page 105, Chapter LXV). As shown in Fig. 2.4 (right), the words '*vortex*' and '*vorticis*' in Latin and the graph of two vortices appear in *Principia Philosophiae* (Descartes 1644, page 107). As an interesting mathematical historical note, Fig. 2.4 (left) also shows the Voronoi diagram as discussed in Liebling and Pourin (2012, page 412, left in Figure 2).

2.4 Physical Studies of *Vortex* by René Descartes in 1633 and 1664

Did Descartes work on the physical studies of *vortex* (Latin)? According to Blackwell's (1966) *Descartes' Laws of Motion*, as early as in 1633, Descartes completed his work entitled *Le monde*. In this work, Descartes presents a theory of motion which ascribes to each body of the universe the power either to remain at rest or to continue in motion in a straight line. Mainly because it advocated a heliocentric astronomy at the height of the Galileo controversy, *Le monde* was finally published posthumously in 1664 (Descartes 1664) as shown in Blackwell (1966, page 220, lines 1–7).

In his interesting article entitled *The Dilemma of Defining a Vortex*, Lugt (1979, page 316, paragraph 1, lines 5–7) finds out that Descartes (1664) already gives the following argument:

> the movement of a body in a fluid necessitates a re-circulatory motion of the fluid which is due to the conservation law of mass in a continuum:

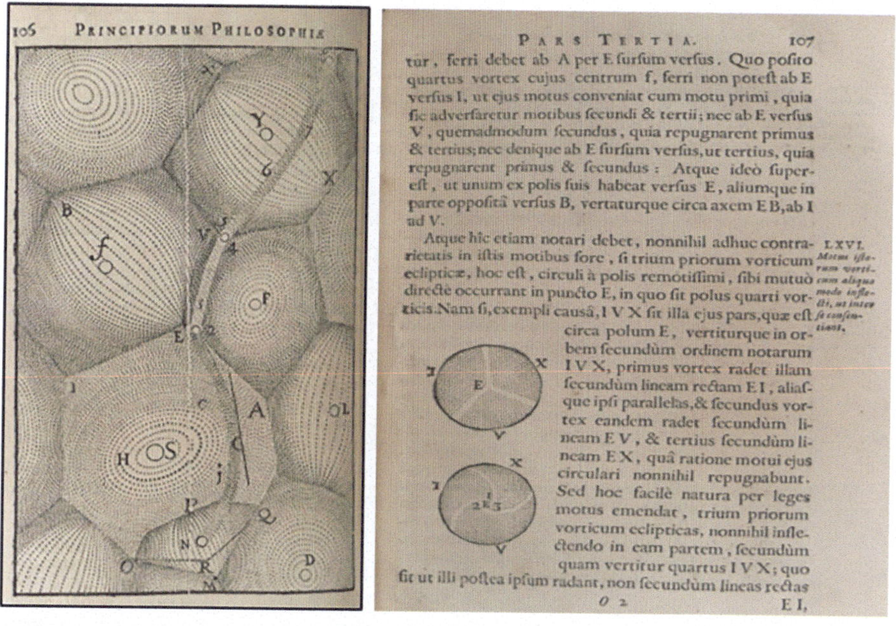

106 PRINCIPIORUM PHILOSOPHIÆ

PARS TERTIA. 107

tur, ferri debet ab A per E ſurſum verſus. Quo poſito quartus vortex cujus centrum f, ferri non poteſt ab E verſus I, ut ejus motus conveniat cum motu primi, quia ſic adverſaretur motibus ſecundi & tertii; nec ab E verſus V, quemadmodum ſecundus, quia repugnarent primus & tertius; nec denique ab E ſurſum verſus, ut tertius, quia repugnarent primus & ſecundus: Atque ideò ſupereſt, ut unum ex polis ſuis habeat verſus E, aliumque in parte oppoſitâ verſus B, vertaturque circa axem EB, ab I ad V.

LXVI. *Motus iſtorum vorticum aliquo modo inflecti, ut inter ſe conſentiant.*

Atque hîc etiam notari debet, nonnihil adhuc contrarietatis in iſtis motibus fore, ſi trium priorum vorticum eclipticæ, hoc eſt, circuli à polis remotiſſimi, ſibi mutuò directè occurrant in puncto E, in quo ſit polus quarti vorticis. Nam ſi, exempli causâ, I V X ſit illa ejus pars, quæ eſt circa polum E, vertiturque in orbem ſecundùm ordinem notarum I V X, primus vortex radet illam ſecundùm lineam rectam E I, aliaſque ipſi parallelas, & ſecundus vortex eandem radet ſecundùm lineam E V, & tertius ſecundùm lineam E X, quâ ratione motui ejus circulari nonnihil repugnabunt. Sed hoc facilè natura per leges motus emendat, trium priorum vorticum eclipticas, nonnihil inflectendo in eam partem, ſecundùm quam vertitur quartus I V X; quo fit ut illi poſtea ipſum radant, non ſecundùm lineas rectas

O 2 E I,

Fig. 2.4 Left: a selected figure showing the poles of each celestial vortex touch the parts of other vortices which are remote from their poles (Descartes 1644, page 106). Right: a selected page showing the appearances of the words '*vortex*' and '*vorticis*' in Latin and the graph of two vortices in *Principia Philosophiae* (Descartes 1644, page 107). (Both are reprinted from the Lessing J. Rosenwald Collection (Library of Congress). Public domain)

This may be the physical studies of *vortex* (Latin) by Descartes before his *Principia Philosophiae*.

2.5 A Golden Spiral by Isaac Newton in 1687

Isaac Newton (1642–1727) was an English mathematician, physicist, astronomer, alchemist, and theologian (right in Fig. 2.1). Newton's whole academic life, from 1661 to 1696, was spent at Trinity College, Cambridge, first as an undergraduate and then as a Fellow from 1667. His detailed biography can be found in Brewster (1855a, b) while an account of his personality in Keynes (1995) and Ackroyd (2006).

Edmund Halley (1656–1742) was an English scientist. Halley played an important role in the publication of Newton's (1687) *Principia*:

> Halley paid all expenses and corrected the proofs himself, and brought *Philosophiae Naturalis Principia Mathematica* to print in 1687. (https://www.space.com/24682-edmond-halley biography.html)

It is interesting to note that the two additional words, *Naturalis* and *Mathematica*, in the title of Newton (1687), were added to the title of Descartes's (1644) *Principia Philosophiae*.

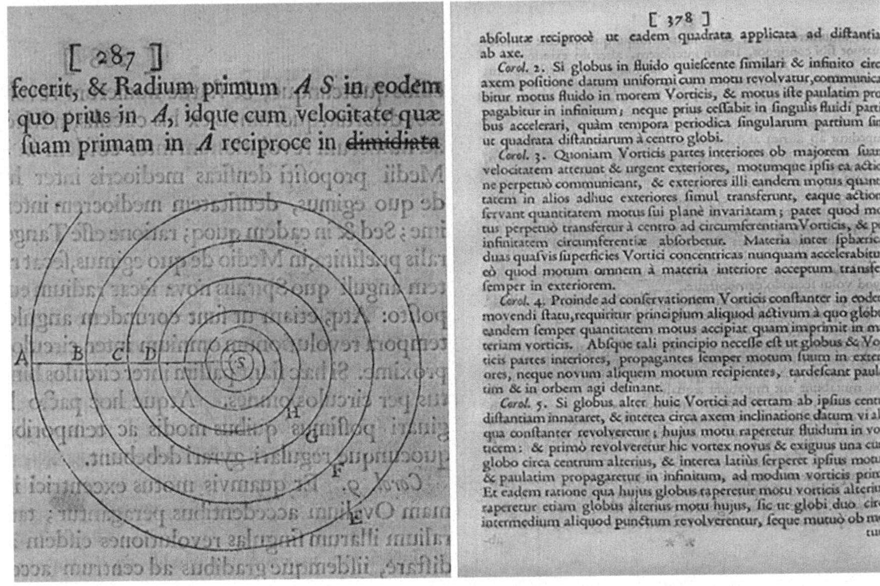

[287]
fecerit, & Radium primum *A S* in eodem
quo prius in *A*, idque cum velocitate quæ
ſuam primam in *A* reciproce in dimidiata

[378]
abſolutæ reciprocè ut eadem quadrata applicata ad diſtantias ab axe.

Corol. 2. Si globus in fluido quieſcente ſimilari & infinito circa axem poſitione datum uniformi cum motu revolvatur, communicabitur motus fluido in morem Vorticis, & motus iſte paulatim propagabitur in infinitum; neque prius ceſſabit in ſingulis fluidi partibus accelerari, quàm tempora periodica ſingularum partium ſint ut quadrata diſtantiarum à centro globi.

Corol. 3. Quoniam Vorticis partes interiores ob majorem ſuam velocitatem atterunt & urgent exteriores, motumque ipſis ea actione perpetuò communicant, & exteriores illi eandem motus quantitatem in alios adhuc exteriores ſimul transferunt, eaque actione ſervant quantitatem motus ſui planè invariatam; patet quod motus perpetuò transfertur à centro ad circumferentiam Vorticis, & per infinitatem circumferentiæ abſorbetur. Materia inter ſphæricas duas quaſvis ſuperficies Vortici concentricas nunquam accelerabitur, eò quod motum omnem à materia interiore acceptum transfert ſemper in exteriorem.

Corol. 4. Proinde ad conſervationem Vorticis conſtanter in eodem movendi ſtatu, requiritur principium aliquod activum à quo globus eandem ſemper quantitatem motus accipiat quam imprimit in materiam vorticis. Abſque tali principio neceſſe eſt ut globus & Vorticis partes interiores, propagantes ſemper motum ſuum in exteriores, neque novum aliquem motum recipientes, tardeſcant paulatim & in orbem agi deſinant.

Corol. 5. Si globus alter huic Vortici ad certam ab ipſius centro diſtantiam innataret, & interea circa axem inclinatione datum vi aliqua conſtanter revolveretur; hujus motu raperetur fluidum in vorticem: & primò revolveretur hic vortex novus & exiguus una cum globo circa centrum alterius, & interea latiùs ſerperet ipſius motus, & paulatim propagaretur in infinitum, ad modum vorticis primi. Et eadem ratione qua hujus globus raperetur motu vorticis alterius, raperetur etiam globus alterius motu hujus, ſic ut globi duo circa intermedium aliquod punctum revolverentur, ſeque mutuò ob motum

Fig. 2.5 Left: a portion of a golden spiral in Newton (1687, page 287). Right: a selected page showing the appearances of the words 'vorticis' and 'vortex' in Latin in Newton (1687, page 378). (Reprinted with permission from Wren Digital Library (Wren Digital Library—NQ.16.200. Accessed on the 27th of November 2025).)

As written in I. Bernard Cohen's A GUIDE TO NEWTON'S PRINCIPIA, "At this time, and for some years to come, Newton was deeply enmeshed in the Cartesian doctrine of vortices. He had no concept of a 'force of gravity acting on the moon in anything like the later sense of the dynamics of the *Principia*.'" (Newton 1999, page 15, lines 5–8). As shown in Fig. 2.5 (left), a golden spiral can be found in Newton (1687, page 287). As shown in Fig. 2.5 (right), the words '*Vorticis*' and '*vortex*' in Latin frequently appear in *Philosophiae Naturalis Principia Mathematica* (Newton 1687, page 378). Can this golden spiral be cautiously interpreted as a spiral vortex?

The answer may be yes. According to note 13. in I. Bernard Cohen's (1914–2003) *A GUIDE TO NEWTON'S PRINCIPIA*, i.e. Hall, Isaac Newton: Adventurer in Thought, p. 62, "Newton had discovered an interesting mathematical correlation within the solar vortex," (Newton 1999, page 15, paragraph 3, lines 1–2).

2.6 *Circulari Fluidorum* and *Vorticis* by Isaac Newton in 1687

In BOOK II OF THE MOTION OF BODIES, Newton studies the motions of fluids at length, including four sections, i.e. SECTION V. *Of the density and compression of fluids; and of hydrostatics* (Newton 1846, First American Edition, pages 293–303); SECTION VII. *Of the motion of fluids, and the resistance made to projected bodies* (Newton 1846, First American Edition, page 323–356); SECTION VIII. *Of motion propagated through fluids* (Newton 1846, First American Edition, pages 356–370); SECTION IX *Of the circular motion of fluids* (Newton 1846, First American Edition, pages 370–379).

Newton (1846, First American Edition, page 293, lines 3–5) gives the definition of a fluid

> A fluid is any body whose parts yield to any force impressed on it, by yielding, are easily moved among themselves.

Notably, the notions 'circular motion', 'angular motions', in particular, and 'vortex' appear numerous times in SECTION IX *Of the circular motion of fluids* (Newton 1846, First American Edition, pages 370–379). As an example shown in Fig. 2.5 (right), the word '*Vorticis*' appears 10 times and '*vortex*' in *Philosophiae Naturalis Principia Mathematica* (Newton 1687, page 378).

Newton first studies the circular motion or the angular motions around a solid cylinder in a uniform and infinite fluid (Newton 1687, page 373; 1846, First American Edition, BOOK II, PROPOSITION LI, pages 370–372) and then a solid sphere in a uniform and infinite fluid (Newton 1687, pages 375–382; 1846, First American Edition, BOOK II, PROPOSITION LII, pages 372–377). Newton's motives for studying fluids, angular motions, circular motion, *vortex* and *vortices* can be seen from his own words (Newton 1687, page 381; 1846, First American Edition, BOOK II, page 377, paragraph 2, lines 1–6):

> I have endeavoured in this Proposition to investigate the properties of vortices, that I might find whether the celestial phenomena can be explained by them; for the phenomenon is this, that the periodic times of the planets revolving about Jupiter are in the sesquiplicate ratio of their distances from Jupiter's centre; the same rule obtains also among the planets that revolve about the sun. And these rules obtain also with the greatest accuracy, as far as has been yet discovered by astronomical observation. Therefore if those planets are carried round in vortices revolving about Jupiter and the sun, the vortices must revolve according to that law.

Like Descartes, Newton assumes the matter of the vortex be fluid. A physical analogy is made between a solid cylinder, a solid sphere, and a planet with regard to the circulation motion or the angular motions. In Newton's own words, "*In this philosophy particular propositions are inferred from the phenomena, and afterwards rendered general by induction*" (Newton 1846, First American Edition, page 507, lines 1–3). Nevertheless, from a historical perspective, most importantly, it should be pointed out that a gravitating force, or a power of gravity, is found to account for the celestial phenomena of the planets and the comets within the universe or the heavens by Newton (1846, First American Edition, BOOK III, pages 391–405, PROPOSITIONS IV, V, VI, VII, VIII, IX, XI, and XIII). The motions of both the planets and the comets can by no means be accounted for by the hypothesis of vortices. At the beginning of GENERAL SCHOLIUM of BOOK III, Newton (1846, page 503, paragraph 2, line 1) writes "The hypothesis of vortices is pressed with many difficulties." As an interesting historical note, like Descartes, as written in GENERAL SCHOLIUM of BOOK III, one can see that Newton also studies nature and the universe with strong religious connotations.

2.7 Discussion

2.7.1 The Significance of Leonardo's Drawing The Deluge

Within the general cultural context, Leonardo's drawing *The Deluge* is a masterpiece created during the last couple of years of his life, when he was living in France at the court of Francis I (reg. 1515–1547). This painting originally depicts the biblical story of the Great Flood, which is a popular theme during the Renaissance time. A description text is at https://www.rct.uk/collection/912380/a-deluge

> A drawing of a dramatic flood in which the atmosphere above a wooded hill has materialised in a gigantic explosion, with jets of water shooting out from the centre. Square blocks of stone topple and fall from the sky. Above is a dark cloud from which jets of rain curl down....Huge cubic blocks of a mountain arch over to crash down at the centre, sending curling waves of debris shooting out like shock-waves to blast the landscape along the lower edge of the sheet.

Is Leonardo's drawing *The Deluge* in 1517 still valid to our pictorial understanding of vortex and turbulence? Leonardo has also been regarded as the first scientist. Within the broader scientific context (e.g. fluid mechanics/fluid dynamo), in a way, Leonardo's drawing *The Deluge* foretells the detailed pictorial structures of a large scale spiral vortex (turbulence) with helicity, which was interpreted by Moffatt and Dormy (2019, page 19, Figure 1.4). We may cautiously say that Leonardo really was 500 years ahead of his time!

We will see the voyage started from Leonardo's drawing *The Deluge* and ended with helicity in 1969 in the following chapters. It has taken more than a few centuries before vortex, vorticity, knot, chirality, and helicity can be mathematically described. There seems to a cyclicity with a period of 500 years.

2.7.2 Are Those Studies by Descartes and Newton Really Invalid?

What is Descartes's motive of studying vortex and vortices in the visible universe? Clearly, Descartes was very ambitious (perhaps visionary), i.e. trying to discover the true nature of the visible universe by using the hypothetical studies of *vortex* (*vortices*).

Are the hypothetical studies of vortex or the hypothesis of vortices by Descartes (1644) and Newton (1687) really invalid to our understandings of the universe? In a broad sense, they are still valid. After Newton's (1687) work on fluid motion, *vortex*, and *vortices*, do others continue to work on them? Based on the literature review above, others do. For example, Thomson (1867a) '*vortex atoms*' is consistent with Descartes's (1644) '*vortex*' (*vortices*) in the universe.

One may wonder why Descartes's definition of the vortex is presented as the early "physical studies". In *Le Monde*, Descartes did consider sufficient the common sense of an educated person and never suggested experimental practices like in Galileo's approach. However, throughout this review, on the one hand, the hypothesis of vortices in the Universe by Descartes (1644) and Newton (1687) is NOT out of date at all, but instead, it has gradually enlightened and stimulated the further mathematical studies of vortex and vorticity; on the other hand, their early hypothesis of this kind also provides a deep insight into the physical picture in the Universe as reconfirmed by modern physicists (e.g. Lugt 1983, 1985). Two interesting citations are given in Lugt (1985):

> The concept of vortices was applied to other areas of science by Democritus, a natural philosopher of the fifth century B.C. who is best known for his notion of the atom; he equated general *physical* laws with vortical motion (Robinson 1968; Lugt 1985, page 162, right column, Vortices, paragraph 1/bottom).

> The next major attempt to formulate a general *physical* theory based on vortices was made by Descartes in 1644 (Aiton 1972). Although this may be considered a forerunner of the field theory, it did not prevail against Newton's mechanics (Lugt 1985, page 162, right column, Vortices, paragraph 2/top).

Note that the word '*physical*' in these lines is highlighted in Italics by the author. Clearly, the early physical studies of vortices were attempted even before Descartes (1644), who stimulated the formulation of a general physical theory based on his studies of vortices, as shown in the first part of this chapter.

As another interesting note, Hans J. Lugt, an American physicist living in modern times, in his article published in *American Scientist*, still repeats or reconfirms the hypothesis of vortices in the Universe by Descartes (1644) and Newton (1687), as shown in the title and subtitle of his paper that "*Vortices and Vorticity in Fluid Dynamics: The development of vortices may serve as a paradigm to illustrate how patterns in nature become organized*" (Lugt 1985).

2.7.3 *The Interdisciplinary Studies of Science and Natural Theology*

Leonardo's drawing *The Deluge* and Descartes's study of nature and the Universe were motivated with strong religious connotations. Both Leonardo and Descartes themselves did not realize that their studies could have stimulated physical and mathematical studies of vortex and helicity. It seems to be true that (1) science and religion are friends and (2) they belong to the one family in the beginning (Shi 2024a, c). Leonardo's drawing *The Deluge* and Descartes's vortex have provided two fine examples of the interdisciplinary studies of science and natural theology/religion.

2.8 Summary

- Pictorial studies of a large scale spiral vortex could originate with Leonardo's drawing *The Deluge* (1517).
- The early hypothetical studies of '*vortex*' ('*vorticis*') within the universe can be found in Descartes (1644, e.g. page 107) and Newton (1687, e.g. page 378).
- The development of vortices mainly by Descartes seems to serve as a paradigm to illustrate how patterns in nature become organized.

Chapter 3
The Mathematical Studies of the Zero-Vorticity Condition of Fluid Motion and Vorticity from 1752 to 1827

If I were again beginning my studies, I would follow the advice of Plato and start with mathematics.

Galileo Galilei (1564–1642)

While Chap. 2 dealt with largely qualitative concepts of "vortices", this one will examine the early stages of the process by which the concept was mathematised—starting, ironically, with theories which required the absence of vorticity.

3.1 The Mathematical Studies of the Zero-Vorticity Condition of Fluid Motion Made by Jean-Baptiste le Rond d'Alembert in 1752

Jean-Baptiste le Rond d'Alembert (1717–1783) was a French mathematician, mechanician, physicist, philosopher, and music theorist (left in Fig. 3.1). How did d'Alembert become interested in fluid motion?

The names M. Newton, M. Jean Bernoulli, M. Daniel Bernoulli, etc., appear frequently throughout in d'Alembert's work on fluid motion (e.g. d'Alembert 1752). Apparently, d'Alembert followed up Newton's (1687) and Bernoulli's (1738) works on fluid motion. For example, d'Alembert (1743) publishes *Traité de dynamique*, which is regarded as a fundamental treatise on dynamics containing the famous "d'Alembert's principle," i.e. Newton's third law of motion (for every action there is an equal and opposite reaction) is true for bodies that are free to move as well as for bodies rigidly fixed. d'Alembert (1744) published *Traité de l'équilibre et du mouvement des fluids*, in which he applied his principle to the theory of equilibrium and motion of fluids. This study led to the development of partial differential equations, a branch of the theory of calculus.

J. Z. Shi, *The Voyage from Vortex to Helicity (1517 – 1969)*, SpringerBriefs in History of Science and Technology,
https://doi.org/10.1007/978-3-032-17853-4_3

Fig. 3.1 From left to right: Jean-Baptiste le Rond d'Alembert (1717–1783). (Reprinted from wikimedia (https://commons.wikimedia.org/wiki/File:Jean_Le_Rond_d%27Alembert,_by_French_school.jpg. Accessed on the 27th of November 2025). Public domain). Leonhard Euler (1707–1783). (Reprinted from wikimedia (https://commons.wikimedia.org/wiki/File:Leonhard_Euler.jpg. Accessed on the 27th of November 2025). Public domain). Joseph-Louis Lagrange (1736–1813). (Reprinted from wikimedia (https://commons.wikimedia.org/wiki/File:Joseph_Louis_Lagrange2.jpg. Accessed on the 27th of November 2025). Public domain). Augustin-Louis Cauchy (1789–1857). (Reprinted from wikimedia (https://commons.wikimedia.org/wiki/File:Cauchy_Augustin_Louis_dibner_coll_SIL14-C2-03a.jpg. Accessed on the 27th of November 2025). Public domain)

Jean-Baptiste le Rond d'Alembert was the early mathematician to describe fluid motion using partial differential equations and a general principle linking statics and dynamics (Darrigol and Frisch 2008, page 1855, Abstract/lines 2–3). For example, d'Alembert (1752, page 104, line 12; page 141, line 16) describes fluid motion with the following partial differential equations:

$$\mu\,\Delta\frac{du}{dt}+\delta\varphi u^{2}=0 \tag{3.1}$$

$$\frac{d(\delta p)}{dz}+\frac{d(\delta q)}{dx}+\frac{\delta p}{z}=0 \tag{3.2}$$

As presented in his *Essai D'une Nouvelle Théorie De La Résistance Des Fluides* (d'Alembert 1752), d'Alembert further studies the resistance of fluid motion. Interestingly,

(a) The following mathematical expression for the zero-vorticity of fluid motion appears in d'Alembert (1752, page 17, line 2 from the bottom; page 195, lines 1 and 2 from the bottom; page 199, lines 3 and 4)

$$\frac{dR}{dy}=\frac{dQ}{dx} \tag{3.3}$$

(b) The following mathematical expression for the zero-vorticity of fluid motion appears in d'Alembert (1752, page 190, line 10; page 191, line 9; page 194, bottom line; page 195, line 1; page 195, line 3 from the bottom; page 196, lines 3 and 13)

$$\frac{d(\delta Q)}{dx} = \frac{d(\delta R)}{dy} \tag{3.4}$$

(c) The following mathematical expression for the zero-vorticity of fluid motion appears in d'Alembert (1752, page 191, line 9; page 195, line 12)

$$\frac{d(Q\delta)}{dx} = \frac{d(R\delta)}{dy} \tag{3.5}$$

(d) The following mathematical expression for the zero-vorticity of fluid motion also appears in d'Alembert (1752, page 100, line 8; page 107, line 11)

$$\frac{dq}{dz} = \frac{dp}{dx} \tag{3.6}$$

The following mathematical expression for the zero-vorticity of fluid motion also appears in d'Alembert (1752, page 195, line 7)

$$\frac{d(Q)}{dx} = \frac{d(R)}{dy} \tag{3.7}$$

As a footnote, the fact that d'Alembert introduced some zero-vorticity condition may not be new, but even so, it will be retained so as to keep this review complete.

3.2 The Mathematical Studies of the Zero-Vorticity Condition of Fluid Motion Made by Leonhard Euler in 1755/1757

Read Euler, read Euler. He is our master in all things.
Dubiously attributed to Laplace by Guglielmo Libri (Sandifer 2010)

Leonhard Euler (1707–1783) was a Swiss mathematician (the second person in Fig. 3.1). Euler is considered to be the greatest mathematician of all time. His interest in fluid motion. Euler (1746) attempts to model light as waves in a frictionless compressible fluid.

As shown in the footnote 36 in Euler (2008, English version, page 1832), '…, Euler recognizes that his previous memoir [Euler 1755a/1757a] on fluid motion was too restricted, in so far as it ignored what we now call vorticity.' This is also evident in Euler (1755a/1757a, page 229, Chapter XXVI, the bottom line) (Fig. 3.2), i.e. the

$$\left(\frac{dL}{dy}\right)=\left(\frac{dM}{dx}\right);\ \left(\frac{dL}{dz}\right)=\left(\frac{dN}{dx}\right),\ \&\ \left(\frac{dM}{dz}\right)=\left(\frac{dN}{dy}\right),$$

Fig. 3.2 A portion of Euler (1755a/1757a, page 229, Chapter XXVI, the bottom line) showing the equality of the cross derivatives of *L*, *M*, and *N* with respect to the coordinates *x*, *y*, and *z*, respectively, i.e. the zero-vorticity condition. (Reprinted with permission from the Euler Archive (https://scholarlycommons.pacific.edu/cgi/viewcontent.cgi?article=1224&context=euler-works. Accessed on the 27th of November 2025))

$$\left(\frac{du}{dy}\right)=\left(\frac{dv}{dx}\right);\ \left(\frac{du}{dz}\right)=\left(\frac{dw}{dx}\right);\ \left(\frac{du}{dt}\right)=\left(\frac{d\Pi}{dx}\right)$$
$$\left(\frac{dv}{dz}\right)=\left(\frac{dw}{dy}\right);\ \left(\frac{dv}{dt}\right)=\left(\frac{d\Pi}{dy}\right);\ \left(\frac{dw}{dt}\right)=\left(\frac{d\Pi}{dz}\right).$$

Fig. 3.3 A portion of Euler (1755b/1757b, page 290, lines 7–8) showing the equality of the cross derivatives of *u*, *v*, *w* and Π with respect to the coordinates *x*, *y*, and *z*, respectively, i.e. the zero-vorticity condition. Also see footnote 33 in Euler's (1755b/1757b) English version (Euler 2008, page 1831). (Reprinted with permission from the Euler Archive (https://scholarlycommons.pacific.edu/cgi/viewcontent.cgi?article=1225&context=euler-works. Accessed on the 27th of November 2025). Public Domain)

$$\left(\frac{du}{dt}\right)=\left(\frac{d\Pi}{dx}\right);\ \left(\frac{dv}{dt}\right)=\left(\frac{d\Pi}{dy}\right);\ \left(\frac{dw}{dt}\right)=\left(\frac{d\Pi}{dz}\right)$$
$$\left(\frac{du}{dy}\right)=\left(\frac{dv}{dx}\right);\ \left(\frac{du}{dz}\right)=\left(\frac{dw}{dx}\right);\ \left(\frac{dv}{dz}\right)=\left(\frac{dw}{dy}\right),$$

Fig. 3.4 A portion of Euler (1755b/1757b, page 314, Chapter LXVI, lines 8–9) showing the equality of the cross derivatives of *u*, *v*, *w* and Π with respect to the coordinates *x*, *y*, and *z*, respectively, i.e. the zero-vorticity condition. (Reprinted with permission from the Euler Archive (https://scholarlycommons.pacific.edu/cgi/viewcontent.cgi?article=1225&context=euler-works. Accessed on the 27th of November 2025). Public Domain)

equalities of the cross derivatives of *L*, *M*, and *N* with respect to the coordinates *x*, *y*, and *z* implying the zero-vorticity condition.

In his paper *Principes généraux de l'etat d'equilibre des fluides*, as shown in Figs. 3.3 and 3.4, Euler (1755b/1757b, page 290; page 314, Chapter LXVI, lines 8–9) writes the equality of the cross derivatives of *u*, *v*, *w* and Π with respect to the coordinates *x*, *y*, and *z* showing the zero-vorticity condition. Also see the footnote 33 in Euler's (1755b/1757b) English version (Euler 2008, page 1831). We can cautiously interpret it as 'zero-vorticity' case of fluid motion. This was called 'a particular one' by Euler (1755b/1757b, page 292, paragraph 1, last line) and Euler (2008, English version, page 1832, paragraph 1, last line).

3.3 The Mathematical Studies of Vorticity Made by Joseph-Louis Lagrange in 1760/1761 and 1781

Joseph-Louis Lagrange (1736–1813) (the third person in Fig. 3.1) was born in Turin, Sarinia-Piedmont (now Italy) on 25 January 1736 and died in Paris, France on 10 April 1813 [as shown at https://mathshistory.st-andrews.ac.uk/Biographies/Lagrange/]. Lagrange excelled in all fields of analysis and number theory and analytical and celestial mechanics.

In his paper *Application de la méthode exposée dans le Mémoire précédent à la solution de différents problèmes de dynamique*, Lagrange J-L (1760/1761, page 442, line 8) proposed the mathematical expressions for the components of the vorticity (μ, ν):

$$\frac{d\alpha}{dy} - \frac{d\beta}{dx} = \mu, \frac{d\alpha}{dz} - \frac{d\gamma}{dx} = \nu,$$

Note that this is not really concerned with fluid motion.

However, as shown in the footnote in Lagrange (1781, page 695), on 22 November 1781, Lagrange (1781) completed/published his 54-page paper in French entitled *Mémoire sur la Théorie du Mouvement des Fluides*. This is concerned with fluid motion only. How did Lagrange become interested in fluid motion? D'Alembert (1743, 1744) studied fluid motion and had developed its mathematical equations. The name, d'Alembert, appears three times in the first paragraph on the first page of Lagrange (1781). Clearly, d'Alembert and Euler have stimulated Lagrange to be interested in fluid motion.

To draw the reader's attention to the merit of Lagrange (1781) which was read by Lagrange on 22 November 1781, Abstract in French in Lagrange (1781, pages 695–696) is now translated into English:

> Since Mr. d'Alembert has reduced the real laws of motion of fluids to simple analytic equations, this approach has become the subject of much work, discussed in the "Opuscules" of Mr. d'Alembert, and in the proceedings of this Academy and in the one from St Petersburg. The general theory has been very much improved in these different pieces of work, but the same cannot be said about the part of the theory related to the application to particular problems. Mr. d'Alembert seems inclined to think that this application is impossible in most cases, especially when considering fluid motion in vases.
>
> After having carefully considered what has been written on the rigorous theory of fluid motion, I tried to lift, or at least to reduce, the difficulties that slowed down the progress of this theory, and forced geometers to satisfy themselves, when solving the simplest prob-

lems, to indirect methods, or to use questionable approximations. This is what motivated my research, which I will present in the following.

Lagrange (1781) includes the following two sections:

Section 1 General consideration on the fundamental equations of fluid motion (pages 696–728)

Section 2: About the motion of homogeneous fluids in the presence of gravity in vases and channels of arbitrary shape (pages 728–748).

Lagrange (1781, page 696, paragraph 3, lines 1–3) writes in French that

…, pour le onseque de chaque particule, les équations
$dx = pdt,\ dy = qdt,\ dz = rdt;$

i.e. in English

… for the motion of each particle, the equations:
$dx = pdt,\ dy = qdt,\ dz = rdt;$

where dx, dy, and dz are the local displacements which effectively define the velocity.

Lagrange (1781, page 708, lines 1–3) writes in French that

13. Considérons d'abord l'équation de la densité trouvée dans le n° 4 pour les onseq compressibles, et supposons

$$\Delta p = \frac{d\alpha}{dt},\quad \Delta q = \frac{d\beta}{dt},\quad \Delta r = \frac{d\gamma}{dt};$$

en regardant les quantités α, β, γ comme des fonctions inconnues de x, y, z t.
i.e. in English

13. Let us consider first the equation for density (Δ), found in item 4 for compressible cases, and let us suppose

$$\Delta p = \frac{d\alpha}{dt},\quad \Delta q = \frac{d\beta}{dt},\quad \Delta r = \frac{d\gamma}{dt};$$

Treating quantities α, β, γ as unknown functions of x, y, z t.
Lagrange does not say "vorticity" because he thinks in terms of "exact differential forms" $pdx + qdy + rdz$ which means $(p, q, r) = \nabla(\varphi)$. Lagrange then discusses the (now) classical cases where vorticity is zero. In particular, he re-derives Bernoulli's equation (in the middle of page 713) (Lagrange 1781, page 713, line 13):

$$\frac{\Pi}{\Delta} = \mathrm{V} - \frac{d\varphi}{dt} - \frac{1}{2}\left(\frac{d\varphi}{dx}\right)^2 - \frac{1}{2}\left(\frac{d\varphi}{dy}\right)^2 - \frac{1}{2}\left(\frac{d\varphi}{dz}\right)^2 \tag{3.8}$$

By analogy to the stress-strain, Lagrange (1781, page 714, line 4) proposes the following three quantities implying the components of vorticity:

$$\alpha = \frac{dp}{dy} - \frac{dq}{dx}, \ \beta = \frac{dp}{dz} - \frac{dr}{dx}, \ \gamma = \frac{dq}{dz} - \frac{dr}{dy};$$

Lagrange (1781, page 714, bottom) further introduces the Greek letters α', β', γ'; α'', β'', γ''; α''', β''', γ''' to stand for the components of vorticity and their mathematical expressions:

$$\alpha' = \frac{dp'}{dy} - \frac{dq'}{dx}, \ \alpha'' = \frac{dp''}{dy} - \frac{dq''}{dx}, \ \alpha''' = \frac{dp'''}{dy} - \frac{dq'''}{dx}$$

$$\beta' = \frac{dp'}{dz} - \frac{dr'}{dx}, \ \beta'' = \frac{dp''}{dz} - \frac{dr''}{dx}, \ \beta''' = \frac{dp'''}{dz} - \frac{dr'''}{dx}$$

$$\gamma' = \frac{dq'}{dz} - \frac{dr'}{dy}, \ \gamma'' = \frac{dq''}{dz} - \frac{dr''}{dy}, \ \gamma''' = \frac{dq'''}{dz} - \frac{dr'''}{dy}$$

Lagrange (1781, page 714, line 4 and bottom) clearly establishes the three components of vorticity and defines it. Lagrange effectively pursues the exploration of the motion when $u \bullet dM$ is a differential, which is only possible when vorticity is zero. This leads him to ask, naturally, how vorticity can appear, if initially zero everywhere. The question is answered at the end of item 19 (Lagrange 1781, page 717, paragraph 2, lines 2 to 5):

As a consequence, if the form

$$pdx + qdy + rdz$$

is not an exact differential at any time, it cannot ever be a differential during the whole motion (Par onsequent, s'il y a un seul instant …, see page 717).
After that, Lagrange wonders how to proceed when ω is non-zero. And sure enough, when ω is non-zero, he seems to start writing the Euler equation in the following form:

$$\frac{du}{dt} + \backslash\omega\backslash\times u = \ldots \tag{3.9}$$

However, Lagrange does not quite go as far as we know. Remarkably, Lagrange identifies vorticity as the source of limitation to the description of fluid motion in terms of potential flows. The mathematical expressions for the Greek letters (α, β, γ) in Lagrange (1781, page 714, line 4) are represented again in his famous book, the first edition of the two-volume *Mécanique Analytique* (Lagrange 1788, page 458, bottom line) and in his revised edition of 1811–1815 (Lagrange 1815, page 307, line 2) (Fig. 3.5).

3.4 The Mathematical Studies of Vorticity Made by Augustin-Louis Cauchy in 1815/1827

August-Louis Cauchy (1789–1857) was a French mathematician (right in Fig. 3.1). Numerous terms in mathematics bear his name: e.g. the Cauchy integral theorem.

As shown in Fig. 3.6, three mathematical expressions for the zero-vorticity condition appear in Cauchy (1815/1827, page 10, bottom):

As shown in Fig. 3.7, three mathematical expressions for the zero-vorticity condition appear in Cauchy (1815/1827, page 14, line 10/equation (8)):

As shown in Fig. 3.8, three mathematical expressions for the zero-vorticity condition appear in Cauchy (1815/1827, page 43, equation (18)):

As shown in Fig. 3.9, three mathematical expressions of the components of vorticity appear in Cauchy (1815/1827, page 41, lines 2–4 from the bottom):

$$\alpha = \frac{dp}{dy} - \frac{dq}{dx}, \quad \beta = \frac{dp}{dz} - \frac{dr}{dx}, \quad \gamma = \frac{dq}{dz} - \frac{dr}{dy};$$

Fig. 3.5 A portion of Lagrange (1815, page 307, line 2) showing three mathematical expressions for the components of vorticity (α, β, γ). (Reprinted from HathiTrust (https://babel.hathitrust.org/cgi/pt?id=hvd.hxj4gl&seq=325&view=1up. Accessed on the 27th of November 2025). Public domain)

$$(3) \quad \frac{du_0}{db} = \frac{dv_0}{da}, \quad \frac{du_0}{dc} = \frac{dw_0}{da}, \quad \frac{dv_0}{dc} = \frac{dw_0}{db}$$

Fig. 3.6 A portion of Cauchy (1815/1827, page 10, §4, line 5/equation (3)) showing three mathematical expressions for the zero-vorticity condition. (Reprinted from HathiTrust (https://babel.hathitrust.org/cgi/pt?id=chi.096424365&seq=26&view=1up. Accessed on the 27th of November 2025). Public domain)

$$(8) \quad \left\{ \frac{du}{dy} = \frac{dv}{dx}, \quad \frac{du}{dz} = \frac{dw}{dx}, \quad \frac{dv}{dz} = \frac{dw}{dy} \right\},$$

Fig. 3.7 A portion of Cauchy (1815/1827, page 14, line 10/equation (8)) showing three mathematical expressions for the zero-vorticity condition. (Reprinted from HathiTrust (https://babel.hathitrust.org/cgi/pt?id=chi.096424365&seq=30&view=1up. Accessed on the 27th of November 2025). Public domain)

$$(18) \qquad \frac{du}{dy} = \frac{dv}{dx}, \quad \frac{dw}{dx} = \frac{du}{dz}, \quad \frac{dv}{dz} = \frac{dw}{dy}.$$

Fig. 3.8 A portion of Cauchy (1815/1827, page 43, equation (18)) showing three mathematical expressions for the zero-vorticity condition. (Reprinted from HathiTrust (https://babel.hathitrust.org/cgi/pt?id=chi.096424365&seq=59&view=1up. Accessed on the 27th of November 2025). Public domain)

$$\frac{du_0}{db} - \frac{dv_0}{da},$$
$$\frac{du_0}{dc} - \frac{dw_0}{da},$$
$$\frac{dv_0}{dc} - \frac{dw_0}{db},$$

Fig. 3.9 A portion of Cauchy (1815/1827, page 41, lines 2–4 from the bottom) showing three mathematical expressions of the components of vorticity. (Reprinted from HathiTrust (https://babel.hathitrust.org/cgi/pt?id=chi.096424365&seq=57&view=1up. Accessed on the 27th of November 2025). Public domain)

$$\frac{du}{dy} - \frac{dv}{dx}, \quad \frac{dw}{dx} - \frac{du}{dz}, \quad \frac{dv}{dz} - \frac{dw}{dy},$$

Fig. 3.10 A portion of Cauchy (1815/1827, page 42, line 2 from the bottom) showing three mathematical expressions for the components of vorticity. (Reprinted from HathiTrust (https://babel.hathitrust.org/cgi/pt?id=chi.096424365&seq=58&view=1up. Accessed on the 27th of November 2025). Public domain)

As shown in Fig. 3.10, three mathematical expressions of the components of vorticity in Cauchy (1815/1827, page 42, line 2 from the bottom):

If we compare the mathematical expressions in Cauchy (1815/1827, page 42, line 2 from the bottom) with the ones in Lagrange (1781, page 714, line 4; 1788, page 458, bottom line; 1815, page 307, line 2), we can argue that Cauchy was inspired by Lagrange. This is also supported by the following two facts:

(a) Based on Cauchy (1815/1827, page 38, paragraph 1 below equation (4)), Lagrange's (1788) *Mecanique analytique* was cited.
(b) Laplace and Lagrange were visitors at the Cauchy family home and Lagrange in particular seems to have taken an interest in young Cauchy's mathematical education, as shown at: https://mathshistory.st-andrews.ac.uk.

According to Truesdell (1954, page 59, paragraph 2), Lagrange (1760) and Cauchy (1815) were the first to introduce single letters to stand for the vorticity components. As shown above, however, the author thinks that Cauchy (1815/1827, page 42, line 2 from the bottom) simply quoted "Lagrange's *Mecanique anlytique* [1. [re] edition, page 453]" as shown in Cauchy (1815/1827, page 38, paragraph 1 below equation (4)). This is also specified in the literature, e.g. Frisch and Villone (2014). According to them, Cauchy did not add anything new to Lagrange's *Mechanique Analytique* but simply quoted Lagrange's work.

According to Truesdell (1954, pages 59–62), Cauchy made a First interpretation: Cauchy's mean value of the rates of rotation of all directions in a plane, and a Second interpretation: Cauchy's mean value of the rates of rotation of perpendicular axes. These seem to be the first geometrical interpretations of the vorticity. We shall see they are important for us to understand what follows in the 1840s.

3.5 Discussion

3.5.1 The Significance of the Time-Derivative of Vorticity in Lagrange (1781)

One may argue that the author seems to disagree with Truesdell (1954). In the first place, the author agreed with Truesdell (1954) in saying that d'Alembert's introduction of velocity potential and streamlines is purely formal. However, as quoted in Sect. 1.2 **Overview**, Truesdell did not just refer to d'Alembert (1749) only, but to others as well, i.e. Euler (1752, 1755a, b), Lagrange (1760), Cauchy (1815/1827). In addition, it is unclear what Truesdell (1954) really meant by saying 'formal'.

If we re-look at Lagrange (1781, page 708, lines 1–3), although Lagrange (1781, page 708, line 3) explicitly treats the quantities α, β, γ as unknown functions of x, y, z t, they alternatively seem to be considering the time-derivative of vorticity. Obviously, Lagrange works with the differential form, i.e. (u_x dx + u_y dy + u_z dz), and focuses on the case where this form is integrable ($d\phi$ = u_x dx + u_y dy + u_z dz), which is effectively saying that $\omega = 0$—and when ω is not equal to 0, of course, it will not work. This was probably the main way of thinking for people at this time. Historically, we would say that the breakthrough came from Clebsch (1859), who "used" the Darboux theorem (for details, see Hardy 2008, page 245, CHAPTER VI, SECT. **129**.), to represent this form u_x dx + u_y dy + u_z dz by $(\lambda d\mu + d\phi)$; and $\omega = \nabla\lambda \times \nabla\mu$. In a sense, this is the conceptual progress after Lagrange. Of course, we know today several limitations about the Clebsch representation; but this is where one goes naturally from what Lagrange had in mind to more realistic flows involving vorticity. According to Bottazzini (1986), Euler and Lagrange were thinking in terms of complete differentials.

3.5.2 *The Interaction Between the Continent and Britain*

From a historical perspective, it is interesting to see the interaction between the Continent and Britain and its importance in the development of the study of fluid motion, vortex, vorticity, knots, helicity, and chirality. For example, natural philosophy originated on the Continent (by Descartes in the Dutch Republic) but greatly influenced Newton in Britain. In return, Newton's study of fluid motion greatly influenced Jean Bernoulli, Daniel Bernoulli, d'Alembert, Euler, Lagrange, and Cauchy on the Continent.

As presented before, in his studies of fluid motion, Euler (1755a/1757a, page 229; 1755b/1757b, page 290, page 314) assumes the zero-vorticity condition and is also aware of its particular case of fluid motion. This might have stimulated Lagrange and naturally led to the non-zero vorticity condition, i.e. the development of the mathematical expression of the vorticity components by Lagrange (1760/1761, page 442, line 8) and of the vorticity components $(\alpha, \ \beta, \ \gamma)$ by Lagrange (1781, page 714, line 4). The mathematical expression of the vorticity components $(\alpha, \ \beta, \ \gamma)$ in Lagrange (1781, page 714, line 4) has indeed stimulated those similar ones in other works later, e.g. Cauchy (1815/1827, page 42, line 2 from the bottom), Hamilton (1824/1828, page 85, line 8), MacCullagh (1839/1846, page 21, equation (c); page 22, line 3 from the bottom; page 23, line 3; page 36, lines 2 and 10 from the bottom); and Thomson (1847, page 63, equation (II)).

3.6 Summary

- It seemed to never enter the mind of anyone before Lagrange to institute the mathematical expression and notion peculiar to the new vortex by which the imagination is freed from the perceptual reference to vortex motion.
- The early mathematical expression for the components of the vorticity should be credited to Lagrange (1781, page 714, line 4) and Lagrange (1788, page 458, bottom line).
- The early physical idea of the vorticity may be rooted in the possible zero-vorticity condition of fluid motion, e.g. d'Alembert (1752) and Euler (1755a/1757a; 1755b/1757b). The early mathematical expressions for the components of the vorticity $(\mu, \ \nu)$ could go back to Lagrange (1760/1761, page 442, line 8) and $(\alpha, \ \beta, \ \gamma)$ to Lagrange (1781, page 714, line 4), in which the time-derivative of vorticity is already considered.

Chapter 4
The Mathematical Studies of the Rays and the Propagation of Light from 1824 to 1846

Beauty is truth, truth beauty—that is all Ye know on earth, and all ye need to know.

John Keats (1795–1821)

4.1 The Mathematical Studies of the Rays of Light Made by William Rowan Hamilton in 1824/1828

William Rowan Hamilton (1805–1865) (left in Fig. 4.1) was an Irish mathematician, physicist, and astronomer. Tributes to him and his biographic memoirs can be found in Whittaker (1954) and Spearman (1995).

Etienne Louis Malus (1775–1812) was a French officer, engineer, physicist, and mathematician. Malus's *Traité d'optique* was of interest to Hamilton. However, Malus appears to Hamilton to have committed some important errors. This motivates Hamilton to study the properties of optical systems. In his first paper *Theory of Systems of Rays*, Hamilton (1824/1828, page 84, equation (G)) writes the following mathematical expressions for the zero-rotary light ray:

$$\frac{d\beta}{dz}-\frac{d\gamma}{dy}=0,\ \frac{d\gamma}{dx}-\frac{d\alpha}{dz}=0,\ \frac{d\alpha}{dy}-\frac{d\beta}{dx}=0 \tag{4.1}$$

Hamilton (1824/1828, page 85, line 8) writes the following three quantities:

$$\frac{d\beta}{dz}-\frac{d\gamma}{dy},\quad \frac{d\gamma}{dx}-\frac{d\alpha}{dz},\quad \frac{d\alpha}{dy}-\frac{d\beta}{dx}$$

which are proportional to (α, β, γ).

J. Z. Shi, *The Voyage from Vortex to Helicity (1517 – 1969)*, SpringerBriefs in History of Science and Technology,
https://doi.org/10.1007/978-3-032-17853-4_4

Fig. 4.1 Left: William Rowan Hamilton (1806–1865). (Reprinted https://www.loc.gov/resource/cph.3c00657/. Accessed on the 27th of November 2025. Public domain). Right: the sculpture of James MacCullagh (1809–1847) by Christopher Moore (1790–1863). (Photo courtesy of the Board of Trinity College Dublin, Ireland. Reprinted with permission. All rights reserved)

4.2 The Mathematical Studies of the Propagation of Light Made by James MacCullagh in 1839/1846

> And I cherish more than anything else the Analogies, my most trustworthy masters. They know all the secrets of Nature…. Kepler [quoted in Polya (1954, page 12)]

James MacCullagh (1809–1847) (right in Fig. 4.1) was an Irish mathematician and theoretical physicist. *The Collected Works of James MacCullagh* can be found in Jellett and Haughton (1880). A detailed account of his life and work can be found in Roget (1847), Scaife (1990), and Spearman (2010). MacCullagh (1839/1846, 1841, 1842) presented a detailed mathematical study of the propagation of light in a biaxial crystal and established the dynamical theory of crystalline reflexion and refraction of light. Can MacCullagh's (1839/1846, 1841, 1842) mathematical works on the propagation of light in a biaxial crystal be related to the mathematics of vorticity in fluid motion? If so, in what way?

The general connection between Hamilton (1824/1828) and MacCullagh (1839/1846, 1841, 1842) can also be found in Darrigol (2010, pages 143 and 149). However, the specific connection, i.e. equation (G) in Hamilton (1824/1828, page 84), is not really presented in Darrigol (2010). James MacCullagh must have read Hamilton (1824/1828, page 84, equation (G); page 85, line 8). As shown in Fig. 4.2, however, MacCullagh (1839/1846, page 21, equation I) presents the mathematical expressions for the three quantities X, Y, Z:

***Lemma* II. Let ξ, η, ζ denote, as before, the displacements of a particle whose initial coordinates are x, y, z; and after putting**

$$\mathrm{x} = \frac{d\eta}{dz} - \frac{d\zeta}{dy}, \quad \mathrm{Y} = \frac{d\zeta}{dx} - \frac{d\xi}{dz}, \quad \mathrm{z} = \frac{d\xi}{dy} - \frac{d\eta}{dx}, \qquad \text{(c)}$$

Fig. 4.2 A portion of MacCullagh (1839/1846, page 21, equation (c)) showing three mathematical expressions for three quantities X, Y, Z. (Reprinted from Hathitrust (https://babel.hathitrust.org/cgi/pt?id=njp.32101079228258&seq=35&view=1up. Accessed on the 27th of November 2025). Public domain)

$$\mathrm{X} = \frac{d\eta}{dz} - \frac{d\zeta}{dy}, \mathrm{Y} = \frac{d\zeta}{dx} - \frac{d\xi}{dz}, \mathrm{Z} = \frac{d\xi}{dy} - \frac{d\eta}{dx}, \tag{4.2}$$

where ξ, η, ζ denote the displacements of a particle.

In addition, four other different letters and the mathematical expression for the similar quantities are presented:

1. MacCullagh (1839/1846, page 22, line 3 from the bottom) writes the following mathematical expressions for the similar three quantities:

$$\mathrm{X}_{,} = \frac{dY}{dz} - \frac{dZ_{,}}{dy}, \mathrm{Y}_{,} = \frac{dZ}{dx} - \frac{dX_{,}}{dz}, \mathrm{Z}_{,} = \frac{dX}{dy} - \frac{dY_{,}}{dx}, \tag{4.3}$$

2. MacCullagh (1839/1846, page 23, line 3) writes the following mathematical expression for the similar three quantities:

$$\mathrm{X}_{,,} = \frac{dY_{,}}{dz} - \frac{dZ_{,}}{dy}, \mathrm{Y}_{,,} = \frac{dZ_{,}}{dx} - \frac{dX_{,}}{dz}, \mathrm{Z}_{,,} = \frac{dX_{,}}{dy} - \frac{dY_{,}}{dx}; \tag{4.4}$$

3. MacCullagh (1839/1846, page 36, line 10 from the bottom) writes the following mathematical expression for the similar three quantities:

$$\mathrm{X}_0' = \frac{d\eta_0'}{dz_0} - \frac{d\zeta_0'}{dy_0}, \mathrm{Y}_0' = \frac{d\zeta_0'}{dx_0} - \frac{d\xi_0'}{dz_0}, \mathrm{Z}_0' = \frac{d\xi_0'}{dy_0} - \frac{d\eta_0'}{dx_0}, \tag{4.5}$$

4. MacCullagh (1839/1846, page 36, line 2 from the bottom) writes the following mathematical expression for the similar three quantities:

$$\mathrm{X}_0'' = \frac{d\eta_0''}{dz_0} - \frac{d\zeta_0''}{dy_0}, \mathrm{Y}_0'' = \frac{d\zeta_0''}{dx_0} - \frac{d\xi_0''}{dz_0}, \mathrm{Z}_0'' = \frac{d\xi_0''}{dy_0} - \frac{d\eta_0''}{dx_0}, \tag{4.6}$$

On 24 May 1841, MacCullagh read *Supplement to a paper on the dynamical theory of crystalline reflexion and refraction*. The mathematical expression of the components of the vorticity in MacCullagh (1839/1846, page 21, equation I) is repeated in MacCullagh (1841, page 97, equation (1)). MacCullagh (1841, page 98, equation (3)) also presents the following mathematical expression for the similar three quantities:

$$\xi = \frac{d\eta_1}{dz} - \frac{d\zeta_1}{dy}, \eta = \frac{d\zeta_1}{dx} - \frac{d\xi_1}{dz}, \zeta = \frac{d\xi_1}{dy} - \frac{d\eta_1}{dx}, \tag{4.7}$$

In his paper entitled *On the dispersion of the optic axes, and of the axes of elasticity, in Biaxial crystals*, MacCullagh (1842, page 295, paragraph 2, lines 8–15) re-iterates MacCullagh (1839/1846) by saying

> I showed that when differentials of the first order only are preserved, the function V—which may perhaps with propriety be called the *potential*, since the motion of the system is potentially, or virtually, included in it—is a function of the second degree, composed of the three quantities X, Y, Z which are connected with the displacements ξ, η, ζ the following relations:

$$X = \frac{d\eta}{dz} - \frac{d\zeta}{dy}, Y = \frac{d\zeta}{dx} - \frac{d\xi}{dz}, Z = \frac{d\xi}{dy} - \frac{d\eta}{dx} \tag{4.8a}$$

MacCullagh (1842, page 296, lines 5–6) re-writes the components of the similar three quantities in MacCullagh (1839/1846, page 22, line 3 from the bottom; page 23, line 3) in the following forms:

$$X_1 = \frac{dY}{dz} - \frac{dZ_{,}}{dy}, Y_1 = \frac{dZ}{dx} - \frac{dX_{,}}{dz}, Z_1 = \frac{dX}{dy} - \frac{dY_{,}}{dx}; \tag{4.8b}$$

$$X_2 = \frac{dY_1}{dz} - \frac{dZ_1}{dy}, Y_2 = \frac{dZ_1}{dx} - \frac{dX_1}{dz}, Z_2 = \frac{dX_1}{dy} - \frac{dY_1}{dx}, \tag{4.8c}$$

What motivated James MacCullagh to use these mathematical expressions for describing the observed phenomenon of light propagation? No doubt, Hamilton's (1824/1828) similar work on the rays of light inspired MacCullagh. In a footnote (MacCullagh 1839/1846, page 26), MacCullagh cited "See the reasoning Lagrange in an analogous case, *Mecanique Analytique*, tom. i. p. 68". It can cautiously be inferred from this that those mathematical expressions of three vorticity components by MacCullagh may be due to Lagrange (1781). However, MacCullagh simply makes use of the mathematical *analogy* of those Greek letters and their mathematical expressions in Lagrange (1781, page 714, line 4). Physically, the observed phenomena of light propagation is analogous to the rotational fluid motion. 'since the motion of the system is potentially, or virtually…' (MacCullagh 1842, page 295, paragraph 2, lines 11–11) implies an analogy between them.

In MacCullagh's (1839/1846) *An essay towards a dynamical theory of crystalline reflexion and refraction*, "he seeks to establish the theory upon a strictly mechanical basis by means of the general dynamical equation of Lagrange" (Jellett and Haughton 1880, page v, lines 13–15).

Spearman (2010, page 119, paragraph 2) summarizes MacCullagh's (1839/1846) achievement:

> MacCullagh's achievement in his 1839 paper was to construct a potential function which, when treated by the laws of analytical dynamics, yielded the appropriate oscillations of the aether particles to describe the observed phenomena of light propagation. It is to some extent a question of semantics whether it is a complete dynamical theory or not.

Based on Jellett and Haughton (1880, page v, lines 13–15) and Spearman (2010, page 119, paragraph 2), we can see that James MacCullagh was inspired by Lagrange. Since the name 'M. Cauchy' appears many times in *The Collected Works of James MacCallugh* (Jellett and Haughton 1880), it can be supposed that MacCullagh was also inspired by Cauchy. From a historical perspective, the Greek letters (ξ, η, ζ) in MacCullagh (1841, page 98, equation (3)) anticipate the Greek letters for the components of vorticity used in Helmholtz (1858, German version, page 31, equation (2)), Helmholtz (1867, English version I of Helmholtz (1858), page 490, lines 2–3 below equation (2)), and von Helmholtz (1978, English version ii of Helmholtz (1858), page 47, equation (2)). Note that MacCullagh read his paper in 1839 and had it published in 1846. For this reason, MacCullagh (1839/1846) was read before Stokes (1845).

Interestingly, MacCullagh (1839/1846, page 21, equation I) also shows, for the first time, the negative of the components of what we now call the curl, namely

$$(\mathrm{X},\mathrm{Y},\mathrm{Z}) = -curl(\xi,\eta,\zeta) \tag{4.9a}$$

Conway (1935/1936, page 25, line 4) gives his following interpretation of MacCullagh (1839/1846, page 21, equation I):

$$(L,M,N) = curl(\xi,\eta,\zeta) \tag{4.9b}$$

Note that there is no negative sign '−'.

If so, MacCullagh (1839/1846, page 21, equation I) was rooted in Lagrange (1781, page 714, line 4).

The equations following MacCullagh (1839/1846, page 21, equation I) show that the curl transforms as a vector [actually as an axial vector] under rotation of axes, something that we now take for granted. MacCullagh (1841, page 97, equation (1)) repeats MacCullagh (1839/1846, page 21, equation I) [as equation (1)] for convenience. That paper generalizes the results of MacCullagh (1839/1846) to include the case where the light waves attenuate, the brightness diminishing as the waves progress.

4.3 Discussion

4.3.1 *Are MacCullagh's Quantities the Components of a Vector?*

There are really two important questions to ask about MacCullagh's (1839/1846, page 42, equation (c)) work:

(a) Where did MacCullagh get the idea to consider the quantities he defines in his equation (c)?
(b) Does it make sense to think of MacCullagh's quantities as the components of a vector, and thus as the first appearance of curl?

Regarding (a), MacCullagh was influenced by Hamilton's "*Theory of Systems of Rays*", and it is clear from MacCullagh's (1839/1846, page 26) footnote that he is familiar with Lagrange.

(b) is a more subtle question. Technically, the concept of a "vector" as an algebraic object was only introduced by Hamilton (1847) as the imaginary part of a quaternion (see Crowe, page 5; at https://sites.math.washington.edu/~morrow/335_18/Crowe-HistoryOfVectorAnalysis.pdf): note that this includes the concept of a vector product which is crucial to how we now understand curl. However, as Crowe also shows, there were already various geometrical ideas which we can now recognise as ancestors of the idea of a vector. MacCullagh's Lemma II seems to be essentially a geometrical result, proving that his "curl" transforms the same way as a geometrical vector—even though he doesn't have the concepts to describe it like that. So it seems that MacCullagh has, in some sense, established the most important point: this particular combination of derivatives behaves in the same way as the components of a force or a line in space (which we would now recognise as vectors).

Regarding the negative sign in Eq. (4.9a), it is worth being consistent with the modern sign convention when we use modern notation—but of course, before there was a geometrical interpretation of curl in terms of rotation, there was no reason why it was important for anyone to have a sign convention.

4.3.2 *What Is the Significance of the Components of Curl?*

What is the significance of the components of curl? Although both the Navier-Stokes equation and Maxwell's equation involve curl in an essential way, it was James MacCullagh who identified the importance of the components of curl (known in North America as Nabla × or $\nabla\times$ and in Eastern Europe as rot). It seems to be quite correct in saying that James MacCullagh was the first to describe what we now call the curl, or rot, of a vector function, well before vector notation appeared and before the introduction of the notion or term "curl" in 1871 by James Clerk Maxwell

(1831–1879). It was MacCullagh who introduced the concept of curl by way of its components. MacCullagh had strong geometrical insight but it is still astounding that he came across an expression involving the components of curl which explained optical phenomena so well, as described in Professor Nigel Buttimore's talk at https://www.maths.tcd.ie/~nhb/j/MacC.php mentioning the contributions of George Green (1793–1841), Bruce Hunt, George Francis Fitzgerald (1851–1901), Olivier Darrigol, Richardson Feynman (1918–1988), and Roger Penrose.

$$\frac{d^2\xi}{dt^2} = c^2\frac{dZ}{dy} - b^2\frac{dY}{dz} \tag{4.10a}$$

$$\frac{d^2\eta}{dt^2} = a^2\frac{dX}{dy} - c^2\frac{dZ}{dx} \tag{4.10b}$$

$$\frac{d^2\zeta}{dt^2} = b^2\frac{dY}{dy} - a^2\frac{dX}{dy} \tag{4.10c}$$

As an interesting historical anecdote, his engineering colleague, Professor Brendan Scaife, told Professor Buttimore that Maxwell did not particularly like the term "curl" and removed it from a later edition of his *Treatise on Electricity and Magnetism* (Maxwell 1873b, page 25). However, "curl" remained in the Index of the book, i.e. "Curl, 25" appears in Maxwell (1873b, page 440) only.

Regarding curl, what is the relation between electromagnetism and fluid dynamics/motion? In Brady and Anderson's (2015) Abstract, their words "In 1861, Maxwell derived two of his equations of electromagnetism by modelling a magnetic line of force as a 'molecular vortex' in a fluid-like medium." seems to be an attempt at answering this question.

4.4 Summary

- Other early mathematical expressions for the components of the vorticity appear in Cauchy (1815/1827), Hamilton (1824/1828), and MacCullagh (1839/1846).
- In that sense, MacCullagh can be given the credit for curl, but it is important to note that he was not developing it out of nowhere—rather, he seems to have been the first person to ask the crucial geometrical questions about it.
- We will see that MaCullugh's mathematical studies of the propagation of light are also significant for the development of chirality as presented in Chap. 9.

Chapter 5
The Mathematical Studies of Fluid Motion, Magnetic and Galvanic Forces, and Vortex Motions from 1845 to 1858

*Physical law should have their mathematical beauty.**

Paul A.M. Dirac (1902–1984)

From a historical perspective, the detailed mathematical studies of fluid motion, which were undertaken by d'Alembert (1752), Euler (1755a/1757a, 1755b/1757b), Lagrange (1781, 1815), and Cauchy (1815/1827), were influencing the British mathematicians directly. For example, Lagrange's (1815) *Mécanique Analytique* has stimulated them and their studies of fluid motion. In his paper entitled *On the Truth of the Hydrodynamical Theorem, that if udx + vdy + wdz be a Complete Differential with respect to x, y, z, at any one instant, it is always so*, Power (1842/1843) has critically scrutinized Lagrange's (1815, Tome II, page 307) work. The three components of vorticity in Lagrange (1781, page 714, line 4; 1815, Tome II, page 307, line 2) also appear in Power (1842/1843, page 457, line 4 from the bottom) and the similar ones in Power (1842/1843, page 459, lines 4–6 from the bottom).

5.1 The Mathematical Studies of Fluid Motion Made by George Gabriel Stokes in 1845

George Gabrial Stokes (1819–1903) was an Irish-English mathematician (left in Fig. 5.1). His work on fluid motion was indeed inspired by Lagrange (1781) and especially Lagrange's (1815) *Mécanique Analytique*. On the one hand, the zero-vorticity of fluid motion appears in Stokes (1842/1843, page 440, line 2 from the bottom; page 449, line 10 from the bottom; page 450, line 7) and Stokes (1843, page 108, equation (C)). On the other hand, in his famous paper *On the theories of the*

*This statement was Dirac's response to the question of his philosophy of physics, posed to him in Moscow on 3 Oct. 1956.

J. Z. Shi, *The Voyage from Vortex to Helicity (1517 – 1969)*, SpringerBriefs in History of Science and Technology,
https://doi.org/10.1007/978-3-032-17853-4_5

Fig. 5.1 From left to right: George Gabriel Stokes (1819–1903), the University of Cambridge, U.K. (Photo by the author. William Thomson (1824–1907). Reprinted from Smithsonian Institution Archives (Smithsonian Institution Archives, Record Unit 95, Box 27E, Image No. SIA_000095_B27E_160). Public domain). Hermann von Helmholtz (1821–1894). (Photo by Lord Kelvin. Reprinted from the James Clerk Maxwell Foundation (https://www.clerkmaxwellfoundation.org/html/pics_contemporaries.html. Accessed on the 27th of November 2025). Public domain). Peter Guthrie Tait (1831–1901). (Reprinted from the Biodiversity Heritage Library (https://www.biodiversitylibrary.org/item/58543#page/7/mode/1upfrontpieces. Accessed on the 27th of November 2025). Public domain)

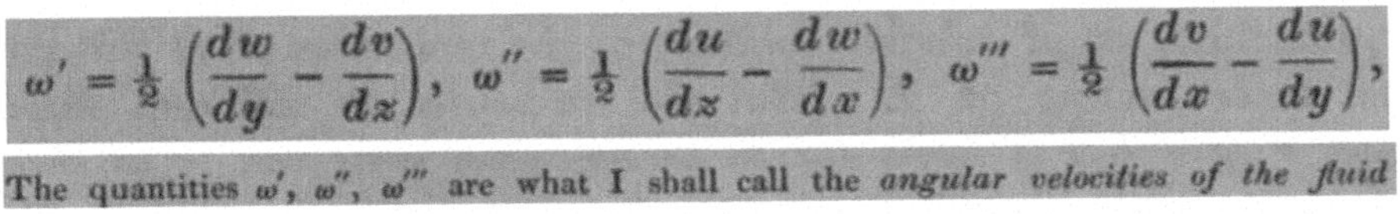

$$\omega' = \frac{1}{2}\left(\frac{dw}{dy} - \frac{dv}{dz}\right), \quad \omega'' = \frac{1}{2}\left(\frac{du}{dz} - \frac{dw}{dx}\right), \quad \omega''' = \frac{1}{2}\left(\frac{dv}{dx} - \frac{du}{dy}\right),$$

The quantities ω', ω'', ω''' are what I shall call the *angular velocities of the fluid*

Fig. 5.2 Top: a portion of Stokes (1845, page 290, equation (1)). Bottom: the appearance of *angular velocities of the fluid*. (Reprinted from the Biodiversity Heritage Library (https://www.biodiversitylibrary.org/item/49441#page/312/mode/1up. Accessed on the 27th of November 2025). Public domain)

internal friction of fluids in motion, and of the equilibrium and motion of elastic solids, Stokes (1845, page 290, equation (1)) presents the following three quantities ω', ω'', ω''' (Fig. 5.2):

$$\omega' = \frac{1}{2}\left(\frac{dw}{dy} - \frac{dv}{dz}\right), \quad \omega'' = \frac{1}{2}\left(\frac{du}{dz} - \frac{dw}{dx}\right), \quad \omega''' = \frac{1}{2}\left(\frac{dv}{dx} - \frac{du}{dy}\right) \tag{5.1}$$

As shown in Fig. 5.2, Stokes (1845, page 290, line 2 from the bottom) writes that "The quantities ω', ω'', ω''' are what I shall call the *angular velocities of the fluid*...". They are represented in Stokes (1845, page 310, line 3).

Who inspired Stokes to propose '*the angular velocities of the fluid*'? Stokes (1845, page 288, paragraph 3) did review Lagrange's proof and M. Cauchy's proof. Cauchy (1815/1827, page 42, line 2 from the bottom) inspired Stokes (1845, page 290, equation (1)). Notably, Stokes adds the factor $\frac{1}{2}$. The factor of $\frac{1}{2}$ is important because it suggests that people were starting to think about vorticity as a physically meaningful quantity (the rate of rotation of a fluid element, or of a small element embedded in the fluid) rather than purely as a mathematical device. This is very much in the spirit of the British school of physical science in that period, which emphasized physical interpretation strongly. It is why the factor of $\frac{1}{2}$ seems natural to Stokes. According to Truesdell (1954, pages 63–64), Stokes makes Third interpretation: Stokes's local angular velocity.

The definitions of $\omega', \omega'', \omega'''$ may first be given in Stokes (1843, page 132, line 2):

> $\omega', \omega'', \omega'''$, the angular velocities of the box about axes through this point parallel to those of x, y, z.

However, their specific mathematical expressions are not given until Stokes (1845, page 290, equation (1)). Just below the equations $\omega' = \frac{1}{2}\left(\frac{dw}{dy} - \frac{dv}{dz}\right)$, $\&c$, Stokes (1845, page 310, paragraph 2) wrote

> We see then that *an indefinitely small element of the fluid*, of which the *three principal moments* about the centre of gravity are equal, if suddenly solidified and detached from the rest of the fluid *will begin to move with* a motion simply of translation, which may however vanish, or *a motion of translation combined with one of rotation*, according as $udx + vdy + wdz$ is, or is not an exact differential, and in the latter case the angular velocities will be the same as in Art. 2.

By 'adding' the factor 1/2, which is obtained after some substitutions (Stokes 1845, pages 306–309), Stokes (1845, page 310, line 3) obtains the equations $\omega' = \frac{1}{2}\left(\frac{dw}{dy} - \frac{dv}{dz}\right)$, $\&c$. It suggests that an indefinitely small element of the fluid will begin to move with a motion of translation combined with one of rotation. Clearly, (1) 'rotation' here must refer to 'vortex' motion; (2) 'strength' of a vortex or 'vortex' motion can be mathematically expressed as $\omega' = \frac{1}{2}\left(\frac{dw}{dy} - \frac{dv}{dz}\right)$, $\&c$. This seems crucial in the change of perception of the concept of vorticity.

5.2 The Mathematical Studies of the Magnetic and Galvanic Forces by William Thomson in 1847

William Thomson (1824–1907) was an Irish-British mathematician, mathematical physicist and engineer (the second person in Fig. 5.1). Following up note 8 in Ranford (2020, page 4, bottom), references to William Thomson (Lord Kelvin) in the reminder of this book will be labelled 'Thomson' (in the period prior to his ennoblement in 1892), 'Kelvin' (subsequently) if appropriate within the general context.

By a mathematical and a physical analogy to the angular velocities of fluid motion in Stokes (1845, page 290, equation (1)), as shown in Fig. 5.3, Thomson (1847, page 63, equation (II)) presents the following halves of expressions the direction $l:m:n$. The components X, Y, Z of the force which this magnet exerts upon an ideal unit of magnetism (one end of a thin uniformly magnetized bar) at the point x, y, z being the differential coefficients of this expression, we have

$$X = \frac{d\beta}{dz} - \frac{d\gamma}{dy}, \quad Y = \frac{d\gamma}{dx} - \frac{d\alpha}{dz}, \quad Z = \frac{d\alpha}{dy} - \frac{d\beta}{dx} \quad ..(\text{II.})$$

The halves of the expressions $\frac{d\beta}{dz} - \frac{d\gamma}{dy}$, &c. indicate the

components round lines parallel to the axes, of the infinitely small rotation which an element of the solid receives, besides its change of form, when $\alpha dx + \beta dy + \gamma dz$ is not a complete differential. This rotation therefore represents the resultant magnetic force, in direction and magnitude.

Fig. 5.3 Top: a portion of Thomson (1847, page 63, equation (II)) showing the components of the vector. (Reprinted from HathiTrust (https://babel.hathitrust.org/cgi/pt?id=hvd.hn46kt&seq=75&view=1up. Accessed on the 27th of November 2025). Public domain). Bottom: a portion of Thomson (1847, page 64, top) explaining the resultant magnetic force in direction and magnitude. (Reprinted from HathiTrust (https://babel.hathitrust.org/cgi/pt?id=hvd.hn46kt&seq=76&view=1up. Accessed on the 27th of November 2025). Public domain)

indicating the components round lines parallel to the axes, of the infinitely small rotation which an element of the solid receives:

$$X = \frac{d\beta}{dz} - \frac{d\gamma}{dy}, \quad Y = \frac{d\gamma}{dx} - \frac{d\alpha}{dz}, \quad Z = \frac{d\alpha}{dy} - \frac{d\beta}{dx} \tag{5.2}$$

This small rotation actually represents the resultant magnetic force in direction and magnitude. Their mathematical expressions seem to be rooted in Lagrange (1781, page 714, line 4).

Just like Stokes (1845, page 290, equation (1)) adds the factor of $\frac{1}{2}$ for the local rotation rate, although Thomson (1847, page 63, equation (II)) gives the expressions $\frac{d\beta}{dz} - \frac{d\gamma}{dy}$, etc., as shown in Fig. 5.3, his comment below the highlighted equation indicates that it is ‘halves’ (i.e. half or $\frac{1}{2}$) of each expression that represents the rotation. This tension between the physically and the mathematically convenient definitions might also explain some of Lamb’s changes in terminology later.

Similarly, Thomson (1847, page 64, equation (III)) writes the following expressions for the components of the galvanic force:

$$\left.\begin{aligned}\frac{d\beta}{dz}-\frac{d\gamma}{dy}=\frac{mz-ny}{r^3}\\ \frac{d\gamma}{dx}-\frac{d\alpha}{dz}=\frac{nx-lz}{r^3}\\ \frac{d\alpha}{dy}-\frac{d\beta}{dx}=\frac{ly-mx}{r^3}\end{aligned}\right\} \tag{5.3}$$

5.3 The Mathematical Studies of Vortex Motions Made by Hermann von Helmholtz in 1858

Hermann von Helmholtz was a German physicist and physician (the third person in Fig. 5.1). There are two biographies available in the English literature (M'Kendrick 1899; Koenigsberger 1906). In his Preface to a biography of Hermann von Helmholtz by Leo Koenigsberger (1906), Lord Kelvin writes

> His [von Helmholtz's] admirable theory of vortex rings is one of the most beautiful of all the beautiful pieces of mathematical work hitherto done in the dynamics of incompressible fluids. Kelvin's Preface in Koenigsberger/Welby (1906)

(a) **The mathematical study of vorticity in Helmholtz** (1858, **German version)**
Helmholtz (1858, German version, page 31, equation (2)) presents the following three relationships:

$$\begin{cases}\dfrac{dv}{dz}-\dfrac{dw}{dy}=2\xi\\ \dfrac{dw}{dx}-\dfrac{du}{dz}=2\eta\\ \dfrac{du}{dy}-\dfrac{dv}{dx}=2\zeta\end{cases} \tag{5.4}$$

"Rotationsbewegungen [ξ, η, ζ]" appears in von Helmholtz (1858, in German, page 31, line 5 below equation (2)) and its English translation may be 'rotational velocities [ξ, η, ζ]'.

Since Helmholtz (1858, German version, page 25, line 6 and footnote*) had read and cited Lagrange (1815, T. II, p. 304), we can infer that Helmholtz (1858, German version, page 31, equation (2)) may have been inspired by Lagrange (1815, page 307, line 2).

(b) **The mathematical study of vorticity in Helmholtz** (1867, **Tait's English version of Helmholtz** (1858))
After the publication of Helmholtz (1858) in German, Peter G. Tait (right in Fig. 5.1) translated it into English (Helmholtz 1867, Tait's English version of Helmholtz (1858)).

From these, by differentiation,

$$\left.\begin{aligned}\frac{dv}{dz}-\frac{dw}{dy}&=2\xi,\\ \frac{dw}{dx}-\frac{du}{dz}&=2\eta,\\ \frac{du}{dy}-\frac{dv}{dx}&=2\zeta.\end{aligned}\right\}\quad\cdots\cdots\quad(2)$$

Fig. 5.4 A portion of Helmholtz (1867, Tait's English version of Helmholtz (1858), page 490, equation (2)) showing three mathematical expressions for the components of vorticity (ξ, η, ζ). (Reprinted with permission from REFERENCE. © 2009, Taylor & Francis (https://www.tandfonline.com/doi/epdf/10.1080/14786446708639824?needAccess=true. Accessed on the 27th of November 2025). All rights reserved)

As shown in Fig. 5.4, Helmholtz (1867, Tait's English version i of Helmholtz (1858), page 490, equation (2)) presented the three relationships as shown in equation (23). Helmholtz (1867, English version of Helmholtz (1858), page 490, lines 2–3 below equation (2)) wrote "the angular velocities [ξ, η, ζ] of fluid element about the coordinate axes."

Tait, the translator of Helmholtz (1858), must have read Stokes (1845) and translated the German word '*Rotationsbewegungen*' (Helmholtz 1858) into the English words '*angular velocities*' used in Stokes (1845), but not '*rotational velocities*'. '*angular velocities*' and '*rotational velocities*' are of the same nature.

(c) **The mathematical study of vorticity in Helmholtz** (1978, **Parpart's English version of Helmholtz** (1858))

Since Tait's translation of Helmholtz (1858) was a rough version, consider the importance of Helmholtz's original precise mathematical treatment of the concepts of vortex lines and vortex filaments in plasma physics, Uwe Parpart translated Helmholtz (1858) into English and claimed his translation to be the first precise rendering into English of Helmholtz's original (Helmholtz 1978, English version of Helmholtz (1858), page 41, bottom paragraph).

Helmholtz (1978, English version ii of Helmholtz (1858, page 46, paragraph 3, line 5) wrote that "the angular velocities are ξ, η, ζ".

Helmholtz (1978, English version ii of Helmholtz (1858, page 47, equation (2)) presented the three relationships as shown in equation (25). Helmholtz (1978, English version ii of Helmholtz (1858, German version, page 47, line 3 below equation (2)) wrote that "the rotational velocities of the water particles around the three coordinate axes". It is unclear why Uwe Parpart, the translator of Helmholtz (1858, German version), used the two different notions for the three quantities.

5.4 Discussion

5.4.1 *What Did Motivate Helmholtz to Become Interested in Vortex Motion?*

There are a few different points of view in regard to this in the literature:

First, based on the footnote 6 in Kragh (2002), Kragh (2002, page 36, paragraph 2) pointed out that Helmholtz's (1858) pioneering work in mathematical hydrodynamics originated in his acoustical research.

Secondly, "His motivations for taking up this new research remain unclear, for at that time Helmholtz was professor of physiology and anatomy at the University of Bonn, and the memoir appeared in the year of his coming to the University of Heidelberg as a professor of physiology" (Meleshko and Aref 2007, page 197, 1. Introduction, lines 5–8).

Thirdly, Leonardo, as a painter, did his extensive pictorial anatomical studies. In a review by Marusic and Broomhall (2021, pages 11–12), Leonardo's pictorial studies of the heart have been related to vortices in fluid mechanics.

By an analogy, we may cautiously infer that Helmholtz's physiological and anatomical work might have motivated him to be interested in the physics of vortex motion and vortices in the sense of the bio-physical or bio-fluid mechanics.

5.4.2 *Impact of Vortex Motion on Modern Fundamental Physics*

As discussed in Kragh (2002, page 93), vortex imagery can be found in modern fundamental physics, e.g. modern particle and field physics, the quantum theory of superfluidity, "skyrmions", non-linear field theory. Vortices are also found in high temperature superconductors (Blatter et al. 1994).

5.5 Summary

- The Greek letters ω', ω'', ω''' in Stokes's (1845) anticipate the modern symbols for the components of vorticity (ξ, η, ζ) in Helmholtz (1858).
- Based on the literature review, those vorticity-related studies done by Stokes, Thomson and Helmholtz seem to be mathematically and physically inspired by Lagrange.
- Fluid motion, the magnetic and galvanic forces, and vortex motions can be mathematically and physically analogous to each other.

Chapter 6
The Mathematical Studies of Molecular Vortices, Vortex Atoms, Spherical Vortex, Spiral Vortex and the Notion 'Vorticity' from 1851 to 1954

There are two types of mind…the mathematical, and what might be called the intuitive.

Blaise Pascal (1623–1662)

This chapter pursues two related developments: (a) the physical interpretation of vorticity in fluid motion; (b) attempts to develop analogous concepts to explain other physical phenomena, ranging from heat to electromagnetism.

6.1 The Hypothesis of Molecular Vortices by William John Macquorn Rankine from 1851 to 1870

William John Macquorn Rankine (1820–1872) was a Scottish engineer and physicist (left in Fig. 6.1). Having been inspired by Sir Humphry Davy and Mr. Joule, Rankine (1851, page 510, Section I./paragraph 1) defines the *hypothesis of molecular vortices* by assuming

> *that each atom of matter consists of a nucleus or central point enveloped by an elastic atmosphere, which is retained in its position by attractive forces, and that the elasticity due to heat arises from the centrifugal force of those atmospheres, revolving or oscillating about their nuclei or central points.*

Having been inspired by Sir Humphry Davy and Mr. Joule, Rankine (1851, page 510, Section I./paragraph 1) defines the *hypothesis of molecular vortices* by assuming

> *that each atom of matter consists of a nucleus or central point enveloped by an elastic atmosphere, which is retained in its position by attractive forces, and that the elasticity due to heat arises from the centrifugal force of those atmospheres, revolving or oscillating about their nuclei or central points.*

J. Z. Shi, *The Voyage from Vortex to Helicity (1517 – 1969)*, SpringerBriefs in History of Science and Technology,
https://doi.org/10.1007/978-3-032-17853-4_6

Fig. 6.1 Left: William John Macquorn Rankine (1820–1872). (Photo by Thomas Annan (1829–1887). Reprinted from Getty Museum Collection (https://www.getty.edu/art/collection/object/108VF4. Accessed on the 27th of November 2025). Public domain). Center: James Clerk Maxwell (1831–1879). (Reprinted from the James Clerk Maxwell Foundation (https://www.clerk-maxwellfoundation.org/html/pics_maxwell.html. Accessed on the 27th of November 2025). Public domain). Right: Horace Lamb (1849–1934). (Reprinted from wikisource (https://en.wikisource.org/wiki/Author:Horace_Lamb. Accessed on the 27th of November 2025). Public domain)

Rankine further applies mathematical analysis to its development. The word "vortex" appears seven times in Rankine (1851). Rankine (1855a, b) studies the hypothesis of molecular vortices, or centrifugal theory of elasticity, and its connexion with the theory of heat. Rankine (1870, page 211, paragraph 2, lines 1–3) defines that 'A vortex, in the most general sense of the word, is a stream or current which circulates within a limited space.'

Rankine (1870, page 220, line 15) writes the following set of partial differential equations:

$$\frac{d\phi}{dx} = \frac{dX}{d\tau}; \frac{d\phi}{dy} = \frac{dY}{d\tau}; \frac{d\phi}{dz} = \frac{dZ}{d\tau}; \tag{6.1}$$

Do those equations imply the zero-vorticity? In the author's view, they do.

6.2 The Zero-Vorticity and Molecular Vortices by James Clerk Maxwell from 1861 to 1873

James Clerk Maxwell (1831–1879) was a Scottish mathematical physicist (the middle person in Fig. 6.1). In Maxwell's (1861a) *On physical lines of force. Part 1. The Theory of molecular vortices applied to magnetic phenomena*, Maxwell makes use of the mathematical analogies of the two problems. In his words, "By making use of the conception of currents in a fluid" (Maxwell 1861a, page 162, paragraph 3, lines

3–4), he shows how to draw lines of forces. Interestingly, "vortices or eddies" appears in Maxwell (1861a, page 165, paragraph 2, line 9), "Every vortex is essentially dipolar," in Maxwell (1861a, page 165, paragraph 3, line 3), and "the velocity of rotation" in Maxwell (1861b, page 165, paragraph 5, line 4).

Maxwell (1861b, page 171, equation (9)) defines the *strength of an electric current parallel to z* throughout a unit of area:

$$\frac{1}{4\pi}\left(\frac{d\gamma}{dy}-\frac{d\beta}{dz}\right)=p,\ \frac{1}{4\pi}\left(\frac{d\alpha}{dz}-\frac{d\gamma}{dx}\right)=q,\ \frac{1}{4\pi}\left(\frac{d\beta}{dx}-\frac{d\alpha}{dy}\right)=r \tag{6.2}$$

where p, q, r are the quantity of electric current per unit of area perpendicular to the axes of x, y, and z, respectively.

The following zero-vorticity components appear in Maxwell (1861b, page 173, equation (15)):

$$\frac{d\gamma}{dy}-\frac{d\beta}{dz}=0,\ \frac{d\alpha}{dz}-\frac{d\gamma}{dx}=0,\ \frac{d\beta}{dx}-\frac{d\alpha}{dy}=0 \tag{6.3}$$

Maxwell (1861b, Part II., page 348) writes the following NOTE:

> Since the first part of this paper [Maxwell 1861a] was written, I have seen in Crelle's *Journal* for 1859 [Helmholtz 1858], a paper by Prof. Helmholtz on Fluid Motion, in which he has pointed out that the lines of fluid motion are arranged according to the same laws as the lines of magnetic force, the path of an electric current corresponding to a line of axes of those particles of the fluid which are in a state of rotation. This is an additional instance of a *physical analogy*, the investigation of which may illustrate both electro-magnetism and hydrodynamics.

Having apparently not been influenced by Helmholtz (1858), Maxwell (1861a, b) seems to have applied those components in molecular vortices to magnetic phenomena independently.

Maxwell (1862, page 85) rewrites

$$\left.\begin{aligned} p&=\frac{1}{4\pi}\left(\frac{d\gamma}{dy}-\frac{d\beta}{dz}\right)\\ q&=\frac{1}{4\pi}\left(\frac{d\alpha}{dz}-\frac{d\gamma}{dx}\right)\\ r&=\frac{1}{4\pi}\left(\frac{d\beta}{dx}-\frac{d\alpha}{dy}\right)\end{aligned}\right\} \tag{6.4}$$

Most interestingly, Maxwell (1862) adds the following lines just below the equation above:

> The same mathematical connexion is found between other sets of phenomena in physical sciences. (1) If α, β, γ represent displacements, velocities, or forces, then p, q, r will be

> rotary displacements, velocities of rotation, or moments of couples producing rotation, in the elementary portions of the mass.

In his *A Treatise on Electricity and Magnetism*, Maxwell (1873a, page 25, equation (1)) presents the following components of the second vector:

$$\xi = \frac{dZ}{dy} - \frac{dY}{dz}, \eta = \frac{dX}{dz} - \frac{dZ}{dx}, \zeta = \frac{dY}{dx} - \frac{dX}{dy} \tag{6.5}$$

Maxwell (1873b, page 490, equation (2)) writes the following components of the angular velocity of an element of the medium:

$$\omega_1 = \frac{1}{2}\frac{d}{dt}\left(\frac{d\zeta}{dy} - \frac{d\eta}{dz}\right) \tag{6.6a}$$

$$\omega_2 = \frac{1}{2}\frac{d}{dt}\left(\frac{d\xi}{dz} - \frac{d\zeta}{dx}\right) \tag{6.6b}$$

$$\omega_3 = \frac{1}{2}\frac{d}{dt}\left(\frac{d\eta}{dx} - \frac{d\xi}{dy}\right) \tag{6.6c}$$

where ξ, η, ζ denote the displacements of the medium in the x, y, and z directions, respectively.

By a comparison with Stokes (1845, page 290, equation (1)), the derivatives of the rotary displacements of the medium with respect to time are made in Maxwell (1873b, page 490, equation (2)).

6.3 The Rectangular Components of the Velocity of the Particle Itself by William Thomson in 1867

In his own words, as presented in his paper entitled *On Vortex Atoms*, Thomson (1867a, page 94, lines 1–5) writes that

> After noticing Helmholtz's admirable discovery of the law of vortex motion in a perfect liquid, that is, in a fluid perfectly destitute of viscosity (or fluid motion), the author said that this discovery inevitably suggests the idea that Helmholtz's rings are the only true atoms.

The notion 'the circular vortex ring' appears in Thomson (1867a, page 98, line 8). Thomson (1867a, page 104, 2nd paragraph, page 105, lines 1–8) briefly introduces Professor Tait's plan of exhibiting smoke rings. As an appendix to Professor Tait's translation of Helmholtz's Memoir on Vortex Motion (Helmholtz 1867), Thomson (1867b) finds the translatory velocity of a circular vortex ring. He uses the symbol ω to stand for the angular velocity of the molecular rotation.

Thomson (1869, page 251, paragraph 3, line 4) presents the following three quantities:

$$\xi = \frac{1}{2}\left(\frac{dw}{dy} - \frac{dv}{dz}\right), \eta = \frac{1}{2}\left(\frac{du}{dz} - \frac{dw}{dx}\right), \zeta = \frac{1}{2}\left(\frac{dv}{dx} - \frac{du}{dy}\right) \tag{6.7}$$

Thomson (1869, page 220, footnote) reads Stokes (1845). Thomson (1869, page 245, paragraph 2, lines 7–8) writes that "Thus ξ, η, ζ are three rectangular components of the velocity of the particle itself."

As an interesting note, Kragh (2002) presents his 83-page account of Thomson's (1867a) *Vortex Atom* which is a Victorian theory of everything. In his concluding remarks, Kragh highlights that the vortex atom theory was 'a serious attempt to understand all of physics on the basis of vortices in a fluid ether.'

6.4 The Notion 'Vorticity' by William Thomson in 1875 and 1887

The notion 'vorticity' appears at least 20 times in Thomson (1875/1878). The conceptual/physical definition of vorticity seems to be (Thomson 1875/1878, page 71, paragraph 2/article 24):

> Vorticity will be used to designate in a general way the distribution of molecular rotation in the matter of a vortex. Thus, if we imagine a vortex divided into a number of infinitely thin vortex filaments, the vorticity will be completely given when the volume of each filament and its circulation, or cyclic constant, are given; but the shapes and positions of the filaments must also be given in order that, not only the vorticity, but its distribution, can be regarded as given.

Does 'the distribution of molecular rotation' (Thomson 1875/1878, page 71, paragraph 2/article 24, line 2) mean '*the rotation velocities*' (Helmholtz 1858, in German, page 31, line 5 below equation (2))? If so, it was Thomson (1875/1878) who first interpreted ξ, η, ζ as 'vorticity'. If not, what did Thomson mean by saying 'the distribution of molecular rotation'? Both 'the distribution of molecular rotation' and 'the rotation velocity' can be a vector quantity of the same nature. Thomson (1875/1878, page 71, paragraph 2/article 24) did not mathematically re-define 'the distribution of molecular motion', i.e. 'vorticity', instead, seemed to interpret ξ, η, ζ as 'vorticity'.

The quotation from Thomson (1875/1878, page 71, paragraph 2/article 24) refers to "circulation, or cyclic constant". This is the same as "circulation" in Kelvin's Circulation Theorem (Thomson 1869), i.e. $\oint_C \boldsymbol{u} \bullet d\boldsymbol{s}$, or the integral of \omega\cdot n over a spanning surface of the curve C. Note that $\boldsymbol{u}$ is the velocity vector, and $d\boldsymbol{s}$ is an element along the closed contour. The fact that he is now using this integrated quantity might be why Thomson does not seem to use "vorticity" for the familiar mathematical quantity.

• The vorticity of an infinitesimal volume dv of fluid is the value of $dv\,.\,\varpi/e$, where ϖ is its molecular rotation, and e the ratio of the distance between two of its particles in the axis of rotation at the time considered, to the distance between the same two particles at a particular time of reference. The amount of the vorticity thus defined for any part of a moving fluid depends on the time of reference chosen. Helmholtz's fundamental theorem of vortex motion proves it to be constant throughout all time for every small portion of an inviscid fluid.

Fig. 6.2 A portion of Thomson (1887a, page 463, footnote*) showing the mathematical definition of the vorticity of an infinitesimal volume of fluid. (Reprinted with permission (https://www.tandfonline.com/doi/epdf/10.1080/14786448708628034?needAccess=true. Accessed on the 27th of November 2025))

By a mathematical and a physical analogy to angular velocities of the fluid motion in Stokes (1845), Thomson (1847, page 63, equation (II)) already uses the same quantities representing the resultant magnetic force in direction and magnitude, it is unclear why those quantities are not directly called or interpreted vorticity in Thomson (1875/1878).

In his paper entitled *On the Stability of Steady and of Periodic Fluid Motion*, Thomson (1887a, page 463, footnote*) gives the following mathematical expression of the vorticity of an infinitesimal volume dv of fluid:

$$\text{the vorticity} = dv \cdot \bar{\varpi} \,/\, e \tag{6.8}$$

where $\bar{\varpi}$ is its molecular rotation, and e the ratio of the distance between two of its particles in the axis of rotation at the time considered, to the distance between the same two particles at a particular time of reference (Fig. 6.2).

Thomson had the following findings:

(a) The amount of the vorticity thus defined for any part of a moving fluid depends on the time of reference chosen.
(b) Helmholtz's fundamental theorem of vortex motion proves it to be constant throughout all time for every small portion of an inviscid fluid.

6.5 The Vortices of the Vortices, or Relative Fluid Motion by Henry Augustus Rowland in 1880

Rowland (1880, page 227, paragraph 4) writes that

> The motion of the zero order is ordinary irrotational fluid motion; the first order is vortex motion; the second order is what I have called the *vortices* of the *vortices*, or *relative fluid motion*,—in an element possessing this motion the centre moves faster than the sides; the third order is one vortex filament inside another and rotating in the opposite direction:

Rowland (1880, page 260, paragraph 4, lines 1–3) writes the following quantities:

$$\mathrm{X}_1 = \frac{d\bar{Z}_0}{dy} - \frac{d\bar{Y}_0}{dz}, \tag{6.9a}$$

$$\mathrm{Y}_1 = \frac{d\bar{X}_0}{dz} - \frac{d\bar{Z}_0}{dx}, \tag{6.9b}$$

$$\mathrm{Z}_1 = \frac{d\bar{Y}_0}{dx} - \frac{d\bar{X}_0}{dy}; \tag{6.9c}$$

6.6 The Re-Interpretation of 'Vorticity' by Horace Lamb from 1879 to 1932

> Every work on fluid dynamics is the better for whatever degree of closeness it attains to the style, clarity, and thoroughness of Sir Horace Lamb's *Hydrodynamics*. Serrin (1959, page 126, paragraph 3, lines 5–7)

Horace Lamb (1849–1934) was a great English mathematician and fluid dynamicist (right in Fig. 6.1). His biography can be found in Love and Glazebrook (1935) and Launder (2012). As an interesting historic note, in his obituary of Sir Horace Lamb (Love and Glazebrook 1935, page 377, paragraph 1, lines 5–13), Augustus Edward Hough Love writes that

> When he [Lamb] arrived at the lecture room, a small room, in Trinity College—I went to look at it recently for the sake of old times—it was filled, but with the scholars and other honours men of the second and third years; our coaches had realized that Lamb would have a message for us. And he realized at once that his audience had not come to listen to the elements of hydrodynamics. He had read Helmholtz's memoir in *Crelle's Journal* for 1858 and Thomson's Edinburgh paper (1867) *On Vortex Motion*.

It can be inferred from Lamb (1879, page v, Preface, line 4) that Lamb might have introduced the vortex of Helmholtz (1858) and Thomson (1869) to his students and scholars in Trinity College, Cambridge before the year of 1874. Based on a course of his lectures delivered in Trinity College, Cambridge in 1874, Horace Lamb published *A Treatise on the Mathematical Theory of the Motion of Fluids* (Lamb 1879). In 1895, Lamb changed the title of his *Treatise* into *Hydrodynamics* (Lamb 1895), which was the second edition. It was revised/published four times (Lamb 1906, 1916, 1924, 1932), namely the third edition, fourth edition, fifth edition, and sixth edition.

1. The interpretation of vorticity in Lamb (1879, 1895)

   Lamb (1879, page 32, Chapter III, article 38, line 5) and Lamb (1895, page 33, article 31, line 5) specified that "due to Stokes*". However, instead of Stokes (1845, page 290, equation (1)), the three components of vorticity ξ, η, ζ in

Thomson (1869, page 251, paragraph 3, line 4) are presented in Lamb (1879, page 32, Chapter III, article 38, line 5) and Lamb (1895, page 33, line 3 from the bottom).

In Preface of the second edition, Lamb (1895, page v, paragraph 3, lines 7–12) writes

> some readers may be disappointed to find that the theory of isolated vortices is still given much in the form in which it was left by the earlier researches of von Helmholtz and Lord Kelvin, and little reference is made to the subsequent investigations of J.J. Thomson, W.M. Hicks, and others, in this field.

J.J. Thomson's investigation refers to *A Treatise on the Motion of Vortex Rings, An Essay to which the Adams Prize was Adjusted in 1882*, in the University of Cambridge (Thomson 1883). Thomson (1883, page 3, PART I., §2, lines 4–5) writes that 'the components of the angular velocity of molecular rotation will be denoted by ξ, η, ζ.'

Lamb (1895, page 34, paragraph 4, lines 3–4) writes that "the component angular velocities of the rotation being ξ, η, ζ". Clearly, Lamb adds the word 'component' before 'angular velocities'.

2. The interpretation of vorticity in Lamb (1906)

   The three components of vorticity (ξ, η, ζ) in Thomson (1869, page 251, paragraph 3, line 4) are represented in Lamb (1906, page 28, CHAPTER III. IRROTATIONAL MOTION, below equation (1)/line 3; page 193, CHAPTER VII VORTEX MOTION, equation (1)).
3. The interpretation of vorticity in Lamb (1916)

   In Preface of the fourth edition, Lamb (1916, page vii, lines 1–5) writes

   > The vector (ξ, η, ζ) as now defined may conveniently be called the 'vorticity', the term 'rotation' being used, if required, in its established sense. It may be added that the altered notation is in conformity with the physical analogies already referred to. It is moreover already current in some writings on the present subject.
4. The interpretation of vorticity in Lamb (1924)

   In Preface of the fifth edition, Lamb (1924, page vi, paragraph 3) writes

   > I ventured in the preceding edition to make a further deviation from prevalent usage, by choosing for special designation the vector (ξ, η, ζ) whose components in terms of the velocity (u, v, w) are

$$\frac{\partial w}{\partial y}-\frac{\partial v}{\partial z}, \quad \frac{\partial u}{\partial z}-\frac{\partial w}{\partial x}, \quad \frac{\partial v}{\partial x}-\frac{\partial u}{\partial y}$$

rather than that whose components have *half* these values. This procedure avoids the insertion of an unnecessary factor 2 or $\frac{1}{2}$ in a number of formulae, in particular in the theorem of Stokes (Art. 32) which is the fundamental relation in the present connection. The vector (ξ, η, ζ) as now defined may conveniently be called the 'vorticity', the term 'rotation' being used, if required, in its established sense. It may be added that the altered notation is in conformity with the physical analogies already referred to. It is moreover already current in some writings on the present subject.

In the fifth edition, the Greek letters ξ, η, ζ are used to denote the components of vorticity. The three components of vorticity (ξ, η, ζ) in Thomson (1869, page 251, paragraph 3, line 4) are represented in Lamb (1924, page 29, CHAPTER III IRROTATIONAL MOTION, equation (2)/line 3; page 185, CHAPTER VII VORTEX MOTION, equation (1)).

5. The interpretation of vorticity in Lamb (1932)

However, in Lamb (1932, page 32), the factor $\frac{1}{2}$ was removed from ξ, η, ζ. Lamb (1932, page 32) writes that:

> the component angular velocities of the rotation being $\frac{1}{2}\xi, \frac{1}{2}\eta, \frac{1}{2}\zeta$
>
> ξ, η, ζ may be conveniently called the vorticity of the medium.

There are three interesting major changes in Lamb (1879, 1895, 1924, 1932):

1. The following relationship can be found in Lamb (1879, 1895):

   the component angular velocities of the rotation = the vorticity of the medium

2. The following relationship can be found in Lamb (1932):

   the component angular velocities of the rotation = $\frac{1}{2}$ the vorticity of the medium

3. Lamb (1924, page 29, foot note †, lines 5–6; 1932, page 31, foot note †, lines 4–6):

   ...the definition of ξ, η, ζ adopted in the text avoids the intrusion of an unnecessary factor 2 (or $\frac{1}{2}$ as the case may be) in this and in a whole series of subsequent formulae relating to vortex motion.

Cauchy (1815/1827, page 42, line 2 from the bottom) was re-adopted in Lamb (1932, page 31, the bottom of equation (2)).

Since Stokes (1845) was published before Helmholtz (1858), the author has a minor comment on Lamb (1932, page 31, foot note †, line 2):

> *for* ξ, η, ζ (Helmholtz) or ω', ω'', ω''' (Stokes) *read* ω', ω'', ω''' (Stokes) or ξ, η, ζ (Helmholtz)

Secondly, Lamb (1932, page 32, paragraph 4) writes that "The vector whose components are ξ, η, ζ may conveniently be called the 'vorticity' of the medium at the point (x, y, z)."

Finally, Lamb cautiously used the words 'may conveniently be', i.e. he was unsure whether it is right to call ξ, η, ζ as vorticity or not. Lamb did not mathematically re-define 'vorticity' as presented in Lamb (1906, page 28, CHAPTER III. IRROTATIONAL MOTION, below equation (1)/line 3; page 193, CHAPTER VII VORTEX MOTION, equation (1), Lamb (1916, page vii, lines 1–5), Lamb (1924, page 29, CHAPTER III IRROTATIONAL MOTION, equation (2)/line 3; page 185, CHAPTER VII VORTEX MOTION, equation (1)). Instead, since Sir

Horace Lamb was a student of George Gabrial Stokes', Lamb (1879, page 32, line 3 from the bottom; 1895, page 33, line 3 from the bottom) faithfully presented Stokes (1845, page 290, equation (1)) and replaced Stokes's (1845) Greek letters ω', ω'', ω''', by using Helmholtz's (1858) Greek letters ξ, η, ζ.

6.7 The Mathematical Studies of Spherical Vortex Made by Micaiah John Muller Hill in 1884 and 1894

Micaiah John Muller Hill (1856–1929) was an English mathematician (left in Fig. 6.3). His obituary notice can be found in Filon (1929). John Lighton Synge (1897–1995) was an Irish mathematician (the second person in Fig. 6.3). His biographical memoir can be found in Florides (2008). Lin Chia-Ch'iao (1916–2013) was a Chinese-American applied mathematician (the middle person in Fig. 6.3).

Based on Clebsch's (1859) components of the velocity of a fluid, Hill (1884) studies the motion of fluid, part of which is moving rotationally and part irrotationally. As an extension of Hill (1884), Hill (1894) publishes the solution for his well-known spherical vortex (Filon 1929, page iii, lines 1–2). Hill's (1894) spherical vortex is discussed in Hicks (1899).

Synge and Lin (1943, page 52, Section 3, lines 1–2) interprets that 'Hill's spherical vortex may be described as a vortex sphere advancing with constant velocity through a fluid which is at rest at infinity.' They attempt to develop a statistical theory of homogeneous isotropic turbulence by regarding the turbulent fluctuations as due to Hill's spherical vortices moving about chaotically, i.e. accepting Hill's spherical vortex as the model of an eddy. To this end, they calculate the velocity correlations due to a Hill's vortex as shown in Synge and Lin (1943, page 49, Figure 1).

Fig. 6.3 From left to right: Micaiah John Muller Hill (1856–1929). (Reprinted with permission from the Royal Society Publishing (http://royalsocietypublishing.org/rspa/article-pdf/124/795/i/25891/rspa.1929.0131.pdfpage 3. Accessed on the 27th of November 2025). All rights reserved). John Lighton Synge (1897–1995). (Reprinted from MathsHistory (https://mathshistory.st-andrews.ac.uk/Biographies/Synge/pictdisplay/. Accessed on the 27th of November 2025). Public domain). Lin, C.C. (Chia-Ch'iao) (1916–2013). (Reprinted from AIP (https://repository.aip.org/node/98099. Accessed on the 27th of November 2025). AIP Emilio Segrè Visual Archives, Physics Today Collection). William Mitchinson Hicks (1850–1934). (Courtesy of the University of Sheffield. All rights reserved). Clifford Ambrose Truesdell (1919–2000). (Reprinted from MathsHistory (https://mathshistory.st-andrews.ac.uk/Biographies//Truesdell/pictdisplay/. Accessed on the 27th of November 2025). Public domain)

As an important historical note, Lin introduced Hill's (1894) spherical vortex and Synge and Lin (1943) to the studies of isotropic turbulence by Chinese turbulence researchers in the 1950s (Shi 2021a).

According to Saffman (1997, page 1950), Taylor (1932) and Taylor and Green (1937) make the early attempts at applying vortex dynamics to the studies of turbulence.

6.8 The Mathematical Studies of Vortex Motion Made by William Mitchinson Hicks in 1899

William Mitchinson Hicks (1850–1934) was a British mathematician and physicist (the fourth person in Fig. 6.3). For his biographical detail, see Milner (1934). Hicks had published a number of papers on vortex motion. Among them, Hicks (1899) seems to be the most important. According to Professor Keith Moffatt (Personal Communication 26 September 2025), the paper of Hicks (1899) was a landmark at the time, though not as well known as Hill (1894) (Hill's spherical vortex).

Joseph John Thomson and Micaiah John Muller Hill were the two referees to Hicks (1899) (Fig. 6.4). Thomson felt that there is a very good deal of analysis which he has not attempted to verify and Hicks's paper is an important and interesting contribution to our knowledge of Vortex Motion. Hill gave a critical but constructive review of Hicks (1899).

Simply to take an example of Hicks's (1899, page 34, the first equation) result:

$$\frac{d^2\psi}{dr^2} + \frac{1}{r^2}\frac{d^2\psi}{d\theta^2} - \frac{\cot\theta}{r^2}\frac{d\psi}{d\theta} = \rho^2\mathrm{F} - f\frac{df}{d\psi} \tag{6.10}$$

where F and f are both functions of ψ.

According to Referee Report by Hill (right in Fig. 6.4), $F(\psi)$ refers to vorticity in Hicks (1899). Comment, '*This is a very important result*', appears on Hill's report (right in Fig. 6.4). Hill also compared Hicks's (1899) result with his own work (Hill 1884, 1894).

6.9 The Kinematics of Vorticity by Clifford Ambrose Truesdell in 1954

Clifford Ambrose Truesdell (1919–2000) was an American mathematician (right in Fig. 6.3). Walter Noll (1925–2017), a student of Truesdell's, writes the following reflections on Truesdell (Noll 2003, page 23, paragraph 1, lines 2, 3, 12 and 13):

> Clifford Truesdell was a singularity among all prominent scientist-scholars of the twentieth century. He believed that the pinnacle of civilization had been reached in the 18th

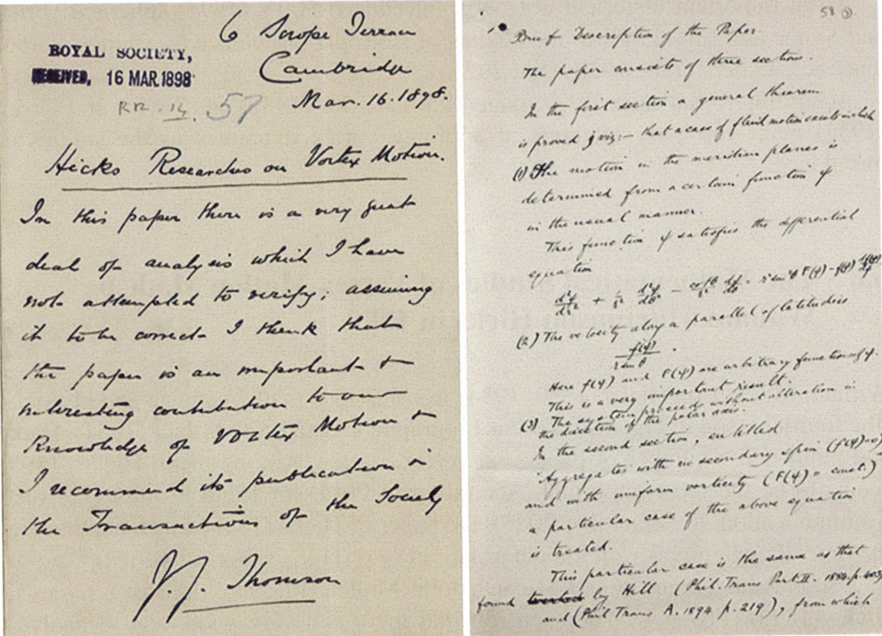

ROYAL SOCIETY,
RECEIVED, 16 MAR 1898
RR. 14. 57

6 Scroope Terrace
Cambridge
Mar. 16. 1898.

Hicks Researches on Vortex Motion.

In this paper there is a very great deal of analysis which I have not attempted to verify; assuming it to be correct I think that the paper is an important & interesting contribution to our knowledge of Vortex Motion & I recommend its publication in the Transactions of the Society

J. J. Thomson

58 (3)

1. Brief Description of the Paper.

The paper consists of three sections.

In the first section a general theorem is proved viz:— that a case of fluid motion exists in which (1) The motion in the meridian planes is determined from a certain function ψ in the usual manner.

This function ψ satisfies the differential equation

(2) The velocity along a parallel of latitude is $\frac{f(\psi)}{r\sin\theta}$.

Here $f(\psi)$ and $F(\psi)$ are arbitrary functions of ψ.

This is a very important result.

(3) The system proceeds without alteration in the direction of the polar axis.

In the second section, entitled "Aggregates with no secondary spin ($f(\psi)=0$) and with uniform vorticity ($F(\psi)=$ const.)" a particular case of the above equation is treated.

This particular case is the same as that found ~~treated~~ by Hill (Phil. Trans Part II. 1884 p. 403) and (Phil Trans A. 1894 p. 219), from which

Fig. 6.4 Left: Referee's report by Joseph John Thomson on Hicks (1899). (Reprinted Joseph John Thomson, Referee's report by Joseph John Thomson, on a paper 'Researches in vortex motion.—Part III. On spiral or gyrostatic vortex aggregates' by William Mitchinson Hicks, 16 March 1898, RR/14/57, The Royal Society Archives, London, https://makingscience.royalsociety.org/items/rr_14_57/referees-report-by-joseph-john-thomson-on-a-paper-researches-in-vortex-motionpart-iii-on-spiral-or-gyrostatic-vortex-aggregates-by-william-mitchinson-hicks?page=1, accessed on 02 December 2025). Right: One page of a 25 page Referee's report by Micaiah John Muller Hill on Hicks (1899). (Reprinted with permission from the Royal Society (Micaiah John Muller Hill, Referee's report by Micaiah John Muller Hill, on a paper 'Researches in vortex motion.—Part III. On spiral or gyrostatic vortex aggregates' by William Mitchinson Hicks, 1898, RR/14/58, The Royal Society Archives, London, https://makingscience.royalsociety.org/items/rr_14_58/referees-report-by-micaiah-john-muller-hill-on-a-paper-researches-in-vortex-motionpart-iii-on-spiral-or-gyrostatic-vortex-aggregates-by-william-mitchinson-hicks?page=3, accessed on 02 December 2025).)

> century…Most importantly, he admired the scientists of the 18th century and, above all, Leonhard Euler, whom he considered to be the greatest mathematician of all time.

In addition, Truesdell was fluent in English, French, German, and Italian (Noll 2003, page 23, paragraph 2, lines 7–8). Not surprisingly, in his book *The Kinematics of Vorticity*, Truesdell (1954) refers to many classic literatures in the 18th century in French and German.

The footnote in Truesdell (1954, page 58) suggests that the name vorticity was introduced by Lamb (1916, **2**, preface & §30). However, from a revisionist perspective, the notion 'vorticity' and its mathematical description first appears in Thomson (1875/1878, page 71, paragraph 2/article 24) and Thomson (1887a, page 463, footnote*).

6.10 Discussion

6.10.1 *What Motivated Rankine to Be Interested in Molecular Vortices?*

As an engineer, Rankine was interested in the elasticity first and then attempted to show "how the laws of the pressure and expansion of gaseous substances may be deduced from that which may be called the *hypothesis of molecular vortices* …that theory which ascribes the elasticity connected with heat to the centrifugal force of small revolutions of the particles of bodies" (Rankine 1851, page 509, paragraph 1).

6.10.2 *The Significance of Vortex Atoms*

Vortex originally had attracted the natural philosophers in the seventeenth century, like DesCartes and Newton, etc. Even in the nineteenth centuries, it is still of interest to natural philosophers. As an example, in his long single review, Kragh (2002) attempts to convince us that the vortex atom is a Victorian theory of everything. The vortex theory was a hydrokinetic theory of matter. According to Thomson (1867a), he said that Helmholtz's admirable discovery of the law of vortex motion in a perfect fluid inevitably suggests the idea that Helmholtz's (1858) rings are the only true atoms. Kragh (2002) argues that the main conceptual advantage of the vortex atom was that it combined atomism with a continuum view of nature.

6.10.3 *Spherical Vortex and Turbulence*

On the one hand, Hill's (1894) spherical vortex stimulated our understanding of isotropic turbulence. On the other hand, ever since Synge and Lin's (1943) work on the application of spherical vortex to turbulence, spherical vortex, vortex rings, and line vortices are found to be the fundamental components of fluid turbulence. This gives a fine example showing how vortex and turbulence are connected with each other.

6.10.4 *Why Is the Notion of Vorticity so Important?*

The notion 'vorticity' by Thomson (1875/1878) is an important physical quantity. The re-interpretation of 'vorticity' by Lamb (1932) has also been significant in our understanding of vortex motion. Lamb changed 'the angular velocities of the rotation' into 'vorticity'.

Why did Lamb change the notion 'the angular velocities of the rotation' into 'vorticity'? Perhaps, Lamb had his physical picture of vortex and its mathematical description. For some unknown reason, Lamb (1932) does not cite Thomson (1875/1878). However, Lamb should have read Thomson (1875/1878) and also was aware of the notion 'vorticity' used in Thomson (1875/1878).

6.10.5 What Is the Significance of Hicks (1899)?

In his Corrigendum to Moffatt (1969), Moffatt (2017) emphasizes that he was unaware of Hicks (1899) in 1969. Based on Moffatt (2017), Hicks vortices refer to the torus knots present among the vortex lines of a family of exact steady axisymmetric solutions of the Euler equations, after Hicks (1899). In a way, we can see how torus knots and vortices are embedded: *the torus knots are present among the vortex lines.*

6.11 Summary

- The early use and conceptual physical definition of the notion 'vorticity' should be credited to Thomson (1875/1878, page 71, paragraph 2/article 24).
- Hill's spherical vortex is useful for our understanding of the physics of isotropic turbulence. Hicks's spiral vortex has stimulated our understanding of knotted vortex.
- Lamb (1932) seems to cautiously re-interpret Stokes's (1845) Greek letters ω', ω'', ω''' and Helmholtz's (1858) Greek letters ξ, η, ζ as vorticity. Following the footsteps of his predecessors, Lamb (1932, page 32, paragraph 4) formally presents his interpretation of the components of the vorticity (ξ, η, ζ) to the fluid mechanics community worldwide.
- The findings from this Chapter are somewhat different from the conventional ones, e.g. those in "Bibliography of vortex dynamics, 1858–1956" by Meleshko and Aref (2007, page 211). Hopefully, this book may be regarded an extension of Truesdell (1954) if any. In particular, it is hoped that the details and findings in this book have further elaborated and clarified those in Truesdell (1954, page 59, paragraph 2).

Chapter 7
The Development of the Mathematical Symbol ∇ from 1831 to 1871

The whole of science is nothing more than a refinement of everyday thinking.

Albert Einstein (1879–1955)

As presented before, along with the physical and mathematical studies of vortex and vorticity and the propagation of light in a biaxial crystal, the curl of a vector function was developed. Interestingly, in a way, the mathematical symbol ∇ (nabla) was also developed. What is the mathematical symbol ∇? How was it developed? Who first developed it? What is the mathematical significance of the mathematical symbol ∇ in vortex and vorticity?

7.1 The Mathematical Symbol ∇ by William Rowan Hamilton from 1831 to 1840

We start our review from the mathematical symbol Δ simply because it already appears in d'Alembert (1743, page 28) and Lagrange (1781, page 697, lines 5–7). However, it is unclear who first uses this mathematical symbol. The mathematical symbol Δ, which applies to the variation of x, is also used numerous times in Herschel (1815/1816). We can cautiously infer that Herschel (1815/1816) may simply have used the same mathematical symbol Δ in d'Alembert (1743, page 28) and Lagrange (1781, page 697, lines 5–7).

Herschel (1815/1816) is referred by Hamilton (1831/1837, page 235, paragraph 1, lines 3–4, paragraph 2). Hamilton (1831/1837) publishes a two-page article entitled *On Differences and Differentials of Zero*. To further develop Herschel (1815/1816), the mathematical symbols Δ and Δ' are considered as marks of differencing (Hamilton 1831/1837, page 235, line 3 from the bottom). From a

J. Z. Shi, *The Voyage from Vortex to Helicity (1517 – 1969)*, SpringerBriefs in History of Science and Technology,
https://doi.org/10.1007/978-3-032-17853-4_7

$$\nabla' f \psi (o') = f(1+\Delta) \nabla' (\psi (o'))^{o}. \qquad \text{(D)}$$

Fig. 7.1 An example of the mathematical symbol ∇′ appearing in Hamilton (1831/1837, page 236, equation (D)). (Reprinted from Hathitrust (https://babel.hathitrust.org/cgi/pt?id=osu.32435072016207&seq=304&view=1up. Accessed on the 27th of November 2025). Public domain)

$$\left.\begin{aligned} \nabla_1 &= \sigma \frac{\delta}{\delta\sigma} + \tau \frac{\delta}{\delta\tau} + \upsilon \frac{\delta}{\delta\upsilon} - 1; \\ \nabla_1' &= \sigma' \frac{\delta}{\delta\sigma'} + \tau' \frac{\delta}{\delta\tau'} + \upsilon' \frac{\delta}{\delta\upsilon'} - 1. \end{aligned}\right\} \quad \text{(M}^{5}\text{)}$$

Fig. 7.2 An example of the mathematical symbol ∇ with the subscript/prime appearing in Hamilton (1832, page 41, equation (M′)). (Reprinted from HathiTrust (https://babel.hathitrust.org/cgi/pt?id=osu.32435072016207&seq=99&view=1up. Accessed on the 27th of November 2025). Public domain)

historical perspective, the mathematical symbol Δ must have further inspired Hamilton to develop the similar mathematical symbols as presented below.

The mathematical symbol ∇′, which is called operation, appears in Hamilton (1831/1837, page 236, equation (D)) (Fig. 7.1).

In his 144 page paper entitled *Third supplement to an essay on the theory of systems of rays*, the mathematical symbol ∇ with the superscript and/or subscript appears in Hamilton (1832, page 41, equation (M′)) is shown in Fig. 7.2

In Fourier's (1822) famous book entitled *Théorie Analytique de la Chaleur* [*The Analytical Theory of Heat* (Fourier 1878)], a very fertile principle, which may be called the *Principle of Fluctuation*, is proposed. It attracted Hamilton's notice and interest. In his own words, to give a new clearness, and even a new extension, to the existing theory of the transformations of arbitrary functions through functions of determined forms, Hamilton (1840/1843) writes his paper entitled *On Fluctuating Functions*. As shown in Fig. 7.3, the mathematical symbol ∇ appears in Hamilton (1840/1843, page 300, equation (s), lines 2, 3 and 5).

The mathematical symbol ∇ appears in Hamilton (1840/1843, page 301, equation (T); page 304, equations (u), (v), and (i^{vu})).

7.2 The Mathematical Symbol ∇ by William Thomson in 1847

In his paper *On a mechanical representation of electric, magnetic, and galvanic forces* (Thomson 1847), Thomson finds three distinct particular solutions of the equations of equilibrium of an elastic solid (Stokes 1845), of which one gives a state

$$f_x = \varpi^{-1} \mathrm{P}_0 \nabla_\infty \int_a^b da\, \mathrm{S}_{a-x,\beta} f_a ; \qquad \text{(S)}$$

which includes the transformations (C) and (R), and in which the notation ∇_∞ is designed to indicate that after performing the operation ∇_β we are to make the variable β infinite, according to some given law of increase, connected with the form of the operation denoted by ∇.

Fig. 7.3 The mathematical symbol ∇ with the subscript and without the subscript appears in Hamilton (1840/1843, page 300, equation (s), lines 2, 3, and 5). (Reprinted from HaithiTrust (https://babel.hathitrust.org/cgi/pt?id=njp.32101079228266&seq=322&view=1up. Accessed on the 27th of November 2025). Public domain)

of the solid in which each element has a certain resultant angular displacement, representing in magnitude and direction of the magnetic force; another represents in a similar rotational manner the galvanic force. Thomson makes a mathematical analogy and a physical analogy to the equilibrium and motion of elastic solids by Stokes (1845), i.e. electric, magnetic, and galvanic forces are analogous to the rotational manner of elastic solids.

In his mathematical solutions, Thomson (1847, page 62, equation (a)) uses the mathematical symbol ∇^2 to denote the operation:

$$\frac{d^2}{dx^2} + \frac{d^2}{dy^2} + \frac{d^2}{dz^2}$$

7.3 The Mathematical Symbol ∇ (Atled) and the Notion 'Curl' by James Clerk Maxwell in 1870

In Maxwell's letter to Tait dated Nov. 7, 1870 sent from Glenlair, Dalbeattie, as shown in Fig. 7.4, we can see that

- The mathematical symbol ∇ appears several times.
- Maxwell was wondering whether the mathematical symbol ∇ should be called 'Atled' or not.
- Maxwell also proposes the notion 'Curl' (line 3 from the bottom).

Following up 'an interesting anecdote' at the end of Sect. 7.5, we can see the above letter in which Maxwell introduces 'Curl' (line 3 from the bottom in Fig. 7.4) before learning that he later decided it was not such a good name and removed it from a later edition of his *Treatise* (Maxwell 1873b, page 25).

GLENLAIR, DALBEATTIE,
Nov. 7, 1870.

Dear Tait

$$\nabla = i\frac{d}{dx} + j\frac{d}{dy} + k\frac{d}{dz}.$$

What do you call this? Atled?

I want to get a name or names for the result of it on scalar or vector functions of the vector of a point.

Here are some rough hewn names. Will you like a good Divinity shape their ends properly so as to make them stick?

(1) The result of ∇ applied to a scalar function might be called the slope of the function. Lamé would call it the differential parameter, but the thing itself is a vector, now slope is a vector word, whereas parameter has, to say the least, a scalar sound.

(2) If the original function is a vector then ∇ applied to it may give two parts. The scalar part I would call the Convergence of the vector function, and the vector part I would call the Twist of the vector function. Here the word twist has nothing to do with a screw or helix. If the word *turn* or *version* would do they would be better than twist, for twist suggests a screw. Twirl is free from the screw notion and is sufficiently racy. Perhaps it is too dynamical for pure mathematicians, so for Cayley's sake I might say Curl (after the fashion of Scroll). Hence the effect of ∇ on a scalar function is to give the slope of that scalar, and its effect on a vector function is to give the convergence and the twirl

Fig. 7.4 A portion of Maxwell's letter to Tait dated Nov. 7, 1870 (Knott 1911, page 143). (Reprinted from the Biodiversity Heritage Library (https://www.biodiversitylibrary.org/page/17116109#page/165/mode/1up. Accessed on the 27th of November 2025). Public domain)

7.4 The Mathematical Symbol ∇ by Peter Guthrie Tait in 1871

At the invitation of Sir Michael Francis Atiyah (1929–2019), Professor H. Keith Moffatt presented a talk entitled *Topological Fluid Dynamics* at Whittaker Colloquium, on 2 October 2017, at the University of Edinburgh, Scotland. As shown at http://www.maths.ed.ac.uk/~aar/moffattslides.pdf, the mathematical symbol ∇ appears on Tait's postcard to Maxwell in 1871 (e.g. slide 7, left column, 2nd postcard, line 4):

$$\int V(d\rho\nabla\sigma) = \int\left(d\rho S\nabla\sigma - \nabla S\sigma d\rho + S(d\rho\nabla)\sigma\right) \tag{7.1}$$

7.5 Discussion

7.5.1 What Is the Significance of the Mathematical Symbol Δ?

Since the mathematical symbol Δ appears in d'Alembert (1743, page 28) and Lagrange (1781, page 697, lines 5–7), the development of the mathematical symbol ∇ may be rooted in d'Alembert (1743, page 28) and Lagrange (1781, page 697, lines 5–7). Important mathematical notions such as ∇ and ∇ × appear in many fields.

Within the general cultural context, the Nabla symbol (∇) has its roots in the Greek alphabet, inspired by the shape of a Phoenician harp, a musical instrument of 10 strings or musical instruments of 12 strings.

However, within the general scientific context, its applications vary across different disciplines, highlighting its importance:

- Mathematics: the Nabla symbol (∇) is crucial in vector calculus, gradients, divergences, and curls
- Physics: the Nabla symbol (∇) is useful in formulating governing equations in electromagnetism and fluid mechanics
- Engineering: the Nabla symbol (∇) represents fields and forces
- Computer Science: the Nabla symbol (∇) is used in algorithms

- (for details, see: https://www.piliapp.com/symbols/nabla/)

7.6 Summary

- The concurrent development of the mathematical concepts of curl and ∇ (nabla) may be related to the mathematical studies of vortex and vorticity.
- Use of the 3-D gradient operator, the mathematical symbol ∇, appears to be rooted in Hamilton (1831/1837) but credited to Hamilton (1840/1843).
- The early mathematical expression of the curl of a vector function should be credited to MacCullagh (1839/1846), in which the components are of opposite sign to those of curl as used today.
- However, the actual notion 'curl' should be credited to James Clerk Maxwell in 1870.

Chapter 8
The Mathematical Studies of Worble and Knot from 1867 to 1885

The beauty of Nature is also reflected in the beauty of natural science.

Werner Karl Heisenberg (1901–1976)

8.1 The Concept of Worble by James Clerk Maxwell in 1867

From a historical perspective, Helmholtz's (1858) theorems of vortex motion stimulated many people's further work at his time, in the chronological order, i.e. worble or worbles by James Maxwell and knots by Peter G. Tait and William Thomson. In fact, worbles and knots are the same.

As shown in Fig. 8.1, in Maxwell's letter to Tait dated 13 November 1867, he kindly requested a spare copy of Tait's translation of Helmholtz on "Water Twists". A copy of its original postcard can be found in Professor Moffatt's talk (slide 3) at http://www.maths.ed.ac.uk/~aar/moffattslides.pdf. Maxwell gradually drew and described the sketches of a simple closed worble, two embracing worbles, and a knotted one.

Maxwell's further discussion with Tait about knotted and linked worbles (vortices) was made in his postcard letter to Tait dated 4 December 1867, as shown in Professor Moffatt's talk (slide 4) at http://www.maths.ed.ac.uk/~aar/moffatt-slides.pdf.

8.2 The Concept of Knot by William Thomson in 1869

His paper entitled *On Vortex Motion* was read by William Thomson on 29th April 1867 (Thomson 1869). It appears to be earlier than Maxwell's letter on November 13, 1867. As Thomson (1869, page 243, line 3 from the bottom) discussed, 'When many rings are linked into one another in various combinations,…', Thomson

J. Z. Shi, *The Voyage from Vortex to Helicity (1517 – 1969)*, SpringerBriefs in History of Science and Technology,
https://doi.org/10.1007/978-3-032-17853-4_8

GLENLAIR,
DALBEATTIE,
Nov. 13, 1867.

Dear Tait

If you have any spare copies of your translation of Helmholtz on "Water Twists" I should be obliged if you could send me one.

I set the Helmholtz dogma to the Senate House in '66, and got it very nearly done by some men, completely as to the calculation, nearly as to the interpretation.

Thomson has set himself to spin the chains of destiny out of a fluid plenum as M. Scott set an eminent person to spin ropes from the sea sand, and I saw you had put your calculus in it too. May you both prosper and disentangle your formulae in proportion as you entangle your worbles. But I fear that the simplest *indivisible* whirl is either two embracing worbles or a worble embracing itself.

For a simple closed worble may be easily split and the parts separated

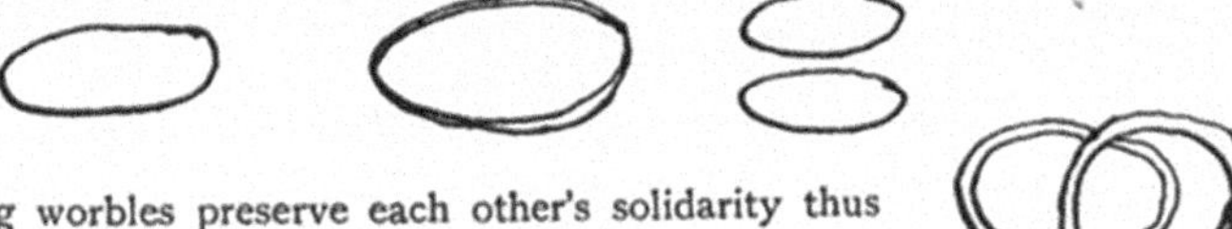

but two embracing worbles preserve each other's solidarity thus

though each may split into many, every one of the one set must embrace every one of the other. So does a knotted one.

yours truly

J. CLERK MAXWELL.

Fig. 8.1 A portion of Knott (1911, page 106) showing 'worble', 'worbles', and 'knotted' in Maxwell's letter to Tait dated on Nov. 13, 1867. (Reprinted from the Biodiversity Heritage Library (https://www.biodiversitylibrary.org/page/17116109#page/129/mode/1up. Accessed on the 27th of November 2025). Public domain)

(1869, page 244, lines 5, 6, 8, 10, graphs of knots) gradually moves onto the term 'knot' and 'knotting', as shown in Fig. 8.2. Among Maxwell, Thomson, and Tait, it is hard to know who first studies the interrelationship between vortex and knot. The early studies of knot may have been done by Thomson (1869, page 244).

8.3 The Concept of Knot by Peter Guthrie Tait from 1877 to 1885

The life and scientific work of Peter Guthrie Tait (1831–1901) can be found in Knott (1911). As presented before, Tait was interested in Helmholtz's (1858) theorems of vortex motion and translated it into its English version (Helmholtz 1867). Furthermore, Tait also did his own experimental illustrations of Helmholtz's (1858)

Fig. 8.2 A portion of Thomson (1869, page 244) showing the knots. (Reprinted from HathiTrust (https://babel.hathitrust.org/cgi/pt?id=mdp.39015017185672&seq=68&view=1up. Accessed on the 27th of November 2025). Public domain)

theorems of vortex motion. According to Knott (1911, page 105, last paragraph, lines 5–7), Tait "was attracted to a study of knots by the problem of the stability of knotted vortex rings such as one might imagine to constitute different types of vortex atoms". Tait communicated a technical note on some elementary properties of closed plane curves, especially with regard to the double points, crossings, or intersections. He pointed out the connection of this subject with the theory of knots, which were discussed in great detail in Tait (1877, 1884, 1885).

By a comparison between Maxwell's worble and Tait's knots in Figs. 8.1 and 8.2, Maxwell in his letter to Tait dated 13 November 1867 expressed explicitly one of the early ideas which Tait finally made the starting point of his study of knots.

The trefoil, the simplest of all knots, was chosen by two Scottish physicists, Balfour Stewart and Tait as a symbolic speculation on the title page of *The Unseen Universe*, which was first published anonymously, though Stewart and Tait acknowledged authorship in the following year. They later published *Paradoxical Philosophy*. A Sequel to *The Unseen Universe* in 1878 (Heimann 1972, page 73, footnote 1). Figure 8.3 (left) shows the trefoil appearing at the top of the title page of Stewart and Tait (1876). A review of knots can be found in Silver (2014).

8.4 Discussion

8.4.1 A Plausible Hypothesis

No doubt, Helmholtz's (1858) theorems of vortex motion (rings) motivated Maxwell's interest in worbles, and Tait's and Thomson's interests in knots. However, if we take the trefoil knot as an example, thinking of its overall shape, there may be

Fig. 8.3 Left: the title page of Stewart and Tait's (1876) *The Unseen Universe* showing the trefoil knot, the simplest of all knots. (Reprinted from HathiTrust (https://babel.hathitrust.org/cgi/pt?id=ia.ark:/13960/s2g2fd4w24m&seq=7&view=1up. Accessed on the 27th of November 2025). Public domain). Right: a picture of *Borromean Rings* taken in Pollokshaws Burgh Hall, Glasgow, Scotland. (Photo by David Pritchard. Reprinted with permission. All rights reserved)

another alternative hypothesis for Maxwell's, Tait's and Thomson's interest in worbles and knots.

As shown in Fig. 8.3 (right), a "*Borromean Rings*" of three interlocking circles is shown on a Glasgow building, Pollokshaws Burgh Hall. In that context it is interesting because most of the symbolism is from Freemasonry, which was very widespread in Scotland at the time. Although neither Maxwell nor Tait and Thomson were involved in it they would have grown up around such details; it could possibly have been a stimulus to their early thought.

8.4.2 *What Is the Mathematical Significance of Knot?*

What is the mathematical significance of knot? The development of knot theory as a recognizable branch of modern topology received considerable stimulus from these investigations (Moffatt 1969, page 118, bottom). The seeds for the development of topology are sowed as a recognizable branch of modern mathematics (Moffatt 2008, page 2, paragraph 1, lines 11–12).

8.4.3 *Implication or Significance of Knots in Fluid Mechanics*

What is the implication or significance of the mathematical study of knots in fluid mechanics? The study of knots mainly is stimulated by Helmholtz's (1858) theorems of vortex motion. It is noteworthy to mention that the study of knots has greatly stimulated the development of topological fluid mechanics (turbulence included). As an example, the *IUTAM Symposium on Topological Fluid Mechanics* was held at

Fig. 8.4 A group photo showing the participants of the IUTAM Symposium on Topological Fluid Mechanics at Cambridge, UK on 13–18 August 1989. (Re-photographed by the author from a poster in Pavilion H at the Department of Applied Mathematics and Theoretical Physics, University of Cambridge, U.K.)

Cambridge UK, 13–18 August 1989 (Fig. 8.4). As the outcome of this symposium, Moffatt and Tsinober (1990) edited its proceedings.

8.4.4 Interrelationship Between Knots and Physics

World Scientific Singapore has been publishing **Series on Knots and Everything**. The title of this series is very striking. As presented in Volume 1 of this series (Kauffmann 2000), knot and link invariants are introduced as generalized amplitudes for quasi-physical processes over an extraordinary range of topics in topology and mathematical physics.

8.5 Summary

- Helmholtz's (1858) theorems of vortex motion do stimulate the developments of knots by Maxwell, Lord Kelvin and Tait.

- The early conceptual studies of knots can be credited to Maxwell's worble in 1867 and Thomson (1869).
- The study of knots has stimulated the developments of a wide range of new topics in physics, including topological fluid mechanics.

Chapter 9
The Conceptual and Mathematical Studies of Chirality from 1856 to 1904

Mystery is God's allurement along the path of knowledge; it is His challenge to a hungry soul.

Maltbie Davenport Babcock (1858–1901)

9.1 The Early Concept of Chirality by William Thomson in 1856 and 1871

In his paper entitled *Dynamical Illustrations of the Magnetic and the Helicoidal Rotatory Effects of Transparent Bodies on Polarized Light*, Thomson (1856, page 151, paragraph 1, line 15) writes the following line:

> …either right-handed or left-handed spirals

Thomson (1856, page 151, paragraph 1, lines 2–3 from the bottom) writes the following line:

> either a right-handed or a left-handed rotation of ordinary light…

These words imply the early concept of chirality arising from the studies of the wave theory of light and lead to the physical and mathematical studies of chirality in the wave theory of light by Thomson himself.

In his paper entitled *On the motion of free solids through a liquid*, Thomson (1871, page 387, lines 1–2) specifies that

> One chief object of this investigation was to illustrate dynamical effects of helicoidal property (that is right or left-handed asymmetry).

Clearly, '…right or left-handed asymmetry' should refer to the early concept of *chirality*.

J. Z. Shi, *The Voyage from Vortex to Helicity (1517 – 1969)*, SpringerBriefs in History of Science and Technology,
https://doi.org/10.1007/978-3-032-17853-4_9

9.2 The Early Concept of Chirality by James Clerk Maxwell in 1873

In his *A Treatise on Electricity and Magnetism*, Maxwell (1873a, pages 24–25) presents a section entitled *On Right-handed and Left-handed Relations in Space*. Maxwell (1873a, page 24, paragraph 2, lines 1–3) points out that "This is the right-handed system which is adopted in Thomson and Tait's *Natural Philosophy*, §243. The opposite, or left-handed system, is adopted in Hamilton's and Tait's *Quaternions*." This also implies the concept of chirality, in particular, it is more explicitly implied in a detailed footnote in Maxwell (1873a, page 24, bottom).

9.3 The Conceptual and Mathematical Studies of Chirality by Lord Kelvin in 1884/1904

Other detailed conceptual, physical, mathematical, and graphical studies of chirality can be found in Lord Kelvin (1884/1904, pages 436–467, Lecture XX; pages 602–661, Appendix H and Appendix I).

9.3.1 The Conceptual Studies of Chirality by Lord Kelvin (1884/1904)

In his Lecture XX, a part of The Wave Theory of Light, which was presented on Friday, October 17, 5 P.M., 1884, Lord Kelvin (1884/1904, page 436) first wonders how well Rankine's old-idea of aerolotropic inertia has served us for the theory of double refraction. The word 'chiral' appears in the following lines (Lord Kelvin 1884/1904, page 436, lines 3–5):

> …a thorough dynamical explanation of the rotation of the plane of polarization of light in a transparent liquid, or crystal, possessing the *chiral* property.

Clearly, in the first place, the word 'chiral' is used by Lord Kelvin to explain chiral polarization of light by aeolotropic inertia or to give a dynamical explanation of a chiral molecule. The word 'chirality' may first appear in Lord Kelvin (1884/1904, page 439, § 204, lines 9 and bottom line):

> …the structure of each molecule has no *chirality*…
> …a molecule which has *chirality*,…

More clearly, in the first place, Lord Kelvin uses the word 'chirality' to describe the structure of a molecule (of the crystal). The word 'chirality' is used through his Lecture XX (e.g. Lord Kelvin 1884/1904, page 461).

9.3.2 *The Mathematical Studies of Chirality by Lord Kelvin (1884/1904)*

Does chirality have any connections with vorticity? Can chirality be expressed by mathematical formulas? If so, what does it really mean?

As shown in Fig. 9.1, to describe the structure of a molecule, Lord Kelvin (1884/1904, page 439, bottom) assumes that ξ, η, ζ and $\dot{\xi}$, $\dot{\eta}$, $\dot{\zeta}$ are the components of displacements and velocity of ether within *B* but not within the part or parts of *B* occupied by the atoms of the molecule. He then obtains the components round *x*, *y*, *z* of rotational velocity (angular velocity) of ether in space in the neighbourhood of *B* and not within any atom. We can see that the components of rotational velocity (angular velocity) are actually vorticities. It is clear that equations (152) in Lord Kelvin (1884/1904, page 439) are consistent with those in Stokes (1845, page 290, equation (1)), Thomson (1847, page 64), Helmholtz (1858, page 31, equation (2)), Helmholtz (1867, page 490, equation (2)), and Thomson (1869, page 251, paragraph 3, line 4).

Based on these components of angular velocity, i.e. vorticities, Lord Kelvin (1884/1904, page 441, equation (158)) derives the mathematical expressions for chirality of virtual inertia as shown in Fig. 9.2.

For the sake of brevity, the detailed derivations of these formulas are not presented here. Interestingly, in a footnote to these formulas, Lord Kelvin emphasizes that the chirality expressed by them is due not to chirality of elasticity but to chirality of virtual inertia.

According to the footnote in Lord Kelvin (1884/1904), equation (158) in Lord Kelvin (1884/1904, page 441) was given by Boussinesq (1903, page 456):

$$\rho\alpha\left(\frac{d\eta''}{dz}-\frac{d\zeta''}{dy}\right),\rho\alpha\left(\frac{d\zeta''}{dx}-\frac{d\xi''}{dz}\right),\rho\alpha\left(\frac{d\xi''}{dy}-\frac{d\eta''}{dx}\right),$$

(§ 201) to each molecule. Let (ξ, η, ζ) and $(\dot{\xi}, \dot{\eta}, \dot{\zeta})$ be the components of displacements and velocity of ether within *B* but not within the part or parts of *B* occupied by the atoms of the molecule. The components round *x*, *y*, *z*, of rotational velocity (commonly, but perhaps less conveniently, called angular velocity) of ether in space in the neighbourhood of *B* and not within any atom, are

$$\frac{1}{2}\left(\frac{d\dot{\zeta}}{dy}-\frac{d\dot{\eta}}{dz}\right);\quad \frac{1}{2}\left(\frac{d\dot{\xi}}{dz}-\frac{d\dot{\zeta}}{dx}\right);\quad \frac{1}{2}\left(\frac{d\dot{\eta}}{dx}-\frac{d\dot{\xi}}{dy}\right)\ \ldots\ldots(152).$$

Fig. 9.1 A portion of Lord Kelvin (1884/1904, page 439, bottom) showing the components of rotational velocity (angular velocity), i.e. vorticities. (Reprinted from HathiTrust (https://babel.hathitrust.org/cgi/pt?id=ien.35556038198842&view=page&seq=469. Accessed on the 27th of November 2025). Public domain)

$$\left.\begin{aligned}
\rho_x\omega^2\xi + \iota\chi_x\frac{\omega^2}{u}(\nu\eta-\mu\zeta) &= (k+\tfrac{1}{3}n)\frac{\omega^2\lambda}{u^2}(\lambda\xi+\mu\eta+\nu\zeta)+n\frac{\omega^2}{u^2}\xi\\
\rho_y\omega^2\eta + \iota\chi_y\frac{\omega^2}{u}(\lambda\zeta-\nu\xi) &= (k+\tfrac{1}{3}n)\frac{\omega^2\mu}{u^2}(\lambda\xi+\mu\eta+\nu\zeta)+n\frac{\omega^2}{u^2}\eta\\
\rho_z\omega^2\zeta + \iota\chi_z\frac{\omega^2}{u}(\mu\xi-\lambda\eta) &= (k+\tfrac{1}{3}n)\frac{\omega^2\nu}{u^2}(\lambda\xi+\mu\eta+\nu\zeta)+n\frac{\omega^2}{u^2}\zeta
\end{aligned}\right\} \quad \text{......(158)*.}$$

Fig. 9.2 A portion of Lord Kelvin (1884/1904, page 441, equation (158)) showing the mathematical expressions for chirality of virtual inertia. Note that χ_x, χ_y, and χ_z are coefficients which are called inertial chiralities of molecule relative to x, y, z, respectively (Lord Kelvin 1884/1904, page 439, bottom 2 lines). (Reprinted from HathiTrust (https://babel.hathitrust.org/cgi/pt?id=ien.35556038198842&view=page&seq=471. Accessed on the 27th of November 2025). Public domain)

where ξ'', η'', ζ'' refer to accelerations while ξ, η, ζ displacements.

After the author's re-checking, no formulas of this nature really appear in Boussinesq (1903, page 456). Those mathematical expressions appearing in Boussinesq (1903, page 456, expressions (218)) are not the same as those in Lord Kelvin (1884/1904, page 441, equation (158)). However, Boussinesq (1903, page 456, expressions (218)) seems to be identical to Lord Kelvin (1884/1904, page 349, expressions (152)). '*rotations moyennes*' appearing in Boussinesq (1903, page 456, bottom line) may be identical to angular velocities.

9.3.3 The Geometrical Study of Opposite Chiralities by Lord Kelvin in 1884/1904

In addition to the quoted physical and mathematical studies, Lord Kelvin moves on to the geometrical study and even graphical representation of the two polarized light waves which are of opposite chiralities: right-handed and left-handed as shown in Fig. 9.3.

As shown in Fig. 9.3, Lord Kelvin (1884/1904, page 446, lines 15–22) describes it in detail:

> The steps of the two spirals (screws) are slightly different, being the spaces travelled by the two waves in their common period; that is to say the wave-lengths of the two waves. To understand this look at fig. 16 showing a right-handed spiral RRR of step 6 cm. and a left-handed spiral LLL of step 5 cm.; having a common axis OX, and both wound on a cylinder of radius 1.5 cm. representing C of (170).

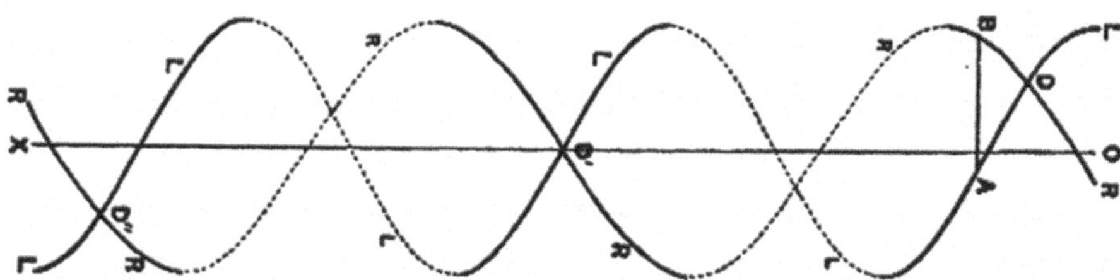

Fig. 9.3 A portion of Lord Kelvin (1884/1904, page 445, Figure 16) showing that the two waves are of opposite chiralities: the former are called right-handed, the latter left-handed. Note that the original graph has been rotated 90° clockwise so as to reduce space in the vertical direction. (Reprinted from HathiTrust (https://babel.hathitrust.org/cgi/pt?id=ien.35556038198842&view=page&seq=475. Accessed on the 27th of November 2025). Public domain)

In a footnote to APPENDIX H. ON THE MOLECULAR TACTICS OF A CRYSTAL, Lord Kelvin (1884/1904, page 619) gives his full conceptual definition of chirality:

> I call any geometrical figure, or group of points, *chiral*, and say that it has chirality if its image in a plane mirror, ideally realized, cannot be brought to coincide with itself. Two equal and similar right hands are homochirally similar. Equal and similar right and left hands are heterochirally similar or 'allochirally' similar (but heterochirally is better). These are also called 'enantiomorphs,' after a usage introduced, I believe, by German writers. Any chiral object and its image in a plane mirror are heterochirally similar.

It should be pointed out that there are two senses of "chirality":

1. As a general synonym for "left- or right-handedness".
2. Specifically as a molecular property associated with certain fluids.

As an interesting historical note, Sir Arthur Stanley Eddington (1882–1944) was a British theoretical physicist and astrophysicist. Based on those at https://archives.trin.cam.ac.uk/index.php/chapter-viii-double-frames, in Chapter VI The Complete Momentum Vector, created in July 1944 and Chapter VIII Double Frames, created in August 1944, of his book manuscript entitled *Fundamental Theory*, Eddington writes the following two chapters:

"§ 55. Equivalence and chirality";
"§ 80. Chirality of a double frame".

This may the early physical and mathematical study of chirality in astrophysics. According to Milne's (1947) book review, Eddington left, almost completed, this manuscript. Sir Edmund Whittaker (1873–1956) was kindly invited by Miss Eddington and the syndics of the Cambridge University Press to supervise the publication of this manuscript in book form (Eddington 1946).

9.4 Discussion

9.4.1 What Motivated William Thomson to Coin 'Chirality'?

From a historical perspective, the author finds out that Thomson's physical and mathematical studies of chirality are rooted in previous studies of the optics in solids and liquids, e.g. MacCullagh (1839/1846). This is evident in Lord Kelvin (1884/1904, page 459).

9.4.2 How Are Knot and Chirality Embedded in Modern Times?

As an important scientific news released on 24 July 2025, a group of Japanese scientists are studying knot and chirality patterns found in both materials and the universe. It was found that the same kind of knotted structures and chiral phenomena are present in both cutting-edge materials studied by materials scientists, and the extreme forms of matter found in the early universe or inside neutron stars studied by high-energy particle physicists. For details, see: https://wpi-skcm2.hiroshima-u.ac.jp/topics/fusion-high-energy-physics/

9.4.3 One of the Acts of Nature Among the Most Diverse Phenomena

If we take a brief survey in the scientific literature, we will be able to find out that chirality is also an important feature in astrophysics and cosmology, and astrobiology, biological systems, chemistry, materials, mathematics, pharmacology, physics, and toxicology. For example,

- The easiest chiral objects to spot around us include horns, shells, spiral staircases, springs, vines and so on.
- Astrophysicists are very interested in magnetic chirality of solar filaments.
- Since all life on Earth is homochiral, earth scientists are involved in their research on terrestrial biochemistry.
- Others include DNA, sucrose and carvone.

Most interestingly, Wiley Periodicals publishes the scientific periodical journal *Chirality*. It is unclear in which field chirality was first studied. Nevertheless, we may say that chirality is indeed one of the acts of nature among the most diverse phenomena.

9.5 Summary

- The helicoidal rotatory effects of transparent bodies on polarized light by Lord Kelvin gradually motivates his own studies of chirality.
- Early conceptual, physical and mathematical studies of chirality can be credited to Thomson (1856) and especially Lord Kevin (1884/1904) in his studies of the wave theory of light.
- It is believed that re-examination of LECTURE XX and other parts of Lord Kelvin (1884/1904) can give us much more information about the development of chirality and its significance.

Chapter 10
The Mathematical Studies of Hélice (French), Helicity, and Schraubensinn (German) from 1811 to 1969

Life is good for only two things, discovering mathematics and teaching mathematics.

Siméon Denis Poisson (1781–1840)

10.1 The Mathematical Study of Hélice (French) by Siméon Denis Poisson in 1811

Siméon Denis Poisson (1781–1840) was a French mathematician (left in Fig. 10.1). In his famous book *Traité de mécanique*, Poisson (1811, pages 503–507) presents the last section entitled *ADDITIONS Loi générale de léquilibre dans les machines*. The word 'hélice' (French) appears many times in Poisson (1811, pages 498, 499, 505).

As shown in Fig. 10.2, for example, the word 'hélice' (French) appears in Poisson (1811, page 505, paragraph 3, lines 3, 9, and 11). The word, 'hélice', is used to describe a geometrical shape like a spiral corkscrew as shown in Poisson (1811, page 512, plate 5, figure 83, figure 84, figure 85). This study can be regarded with caution as one piece of work related to the early simple mathematical study of helicity.

10.2 The Mathematical Studies of Helicoidal Rotatory Effects by William Thomson in 1856

In his paper *Dynamical illustrations of the magnetic and the helicoidal rotatory effects of transparent bodies on polarized light*, Thomson (1856) uses the word 'helicoidal', which might have anticipated the origin of the word '*helicity*'. Does

J. Z. Shi, *The Voyage from Vortex to Helicity (1517 – 1969)*, SpringerBriefs in History of Science and Technology,
https://doi.org/10.1007/978-3-032-17853-4_10

Fig. 10.1 From left to right: Siméon Denis Poisson (1781–1840). (Reprinted from Wikimedia (https://upload.wikimedia.org/wikipedia/commons/b/b7/Simeon_Poisson.jpg. Accessed on the 27th of November 2025). Public domain). Georg Karl Wilhelm Hamel (1877–1954). (Reprinted under CC-BY-SA-4.0 from the University of St Andrews (https://mathshistory.st-andrews.ac.uk/Biographies/Hamel/. Accessed on the 27th of November 2025)). Tsung-Dao Lee (1926–2024). (Photographed by Alan Richards. Reprinted with permission from IAS (https://albert.ias.edu/entities/archivalmaterial/e5a947b3-e20f-485a-8332-180377e8edcb. Accessed on the 27th of November 2025). All rights reserved). Walter M. Elsasser (1904–1991). (Reprinted from AIP (https://repository.aip.org/node/21740. Accessed on the 27th of November 2025). Emilio Segrè Visual Archives, Physics Today Collection). Lodewijk Woltjer (1930–2019). (Reprinted under CC-BY-4.0 license from AAS (https://baas.aas.org/pub/2020i0306/release/1. Accessed on the 27th of November 2025)). Jean Jacques Moreau (1923–2014). (Reprinted under CC-BY-SA-4.0 license from wikiemedia (https://commons.wikimedia.org/wiki/File:Jean-Jacques_Moreau.jpg. Accessed on the 27th of November 2025). Public domain). Robert Betchov (1919–1996). (Courtesy of Renzo Ricca (https://www.renzoricca.com/photo-gallery-1990-2000/. Accessed on the 27th of November 2025). All rights reserved)

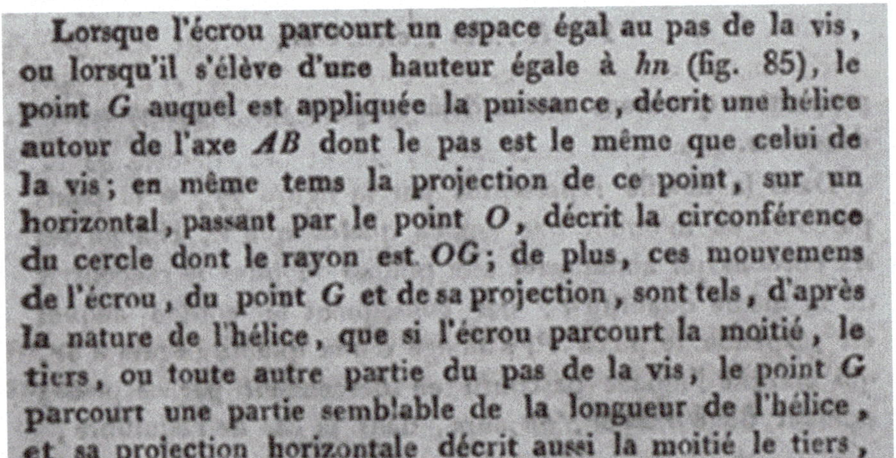

Lorsque l'écrou parcourt un espace égal au pas de la vis, ou lorsqu'il s'élève d'une hauteur égale à *hn* (fig. 85), le point *G* auquel est appliquée la puissance, décrit une hélice autour de l'axe *AB* dont le pas est le même que celui de la vis; en même tems la projection de ce point, sur un horizontal, passant par le point *O*, décrit la circonférence du cercle dont le rayon est *OG*; de plus, ces mouvemens de l'écrou, du point *G* et de sa projection, sont tels, d'après la nature de l'hélice, que si l'écrou parcourt la moitié, le tiers, ou toute autre partie du pas de la vis, le point *G* parcourt une partie semblable de la longueur de l'hélice, et sa projection horizontale décrit aussi la moitié le tiers, ou toute autre partie proportionnelle de la circonférence dont *OG* est le rayon; appelant donc, en général, *r* la quantité dont l'écrou a monté, et *s* l'arc de cercle parcouru

Fig. 10.2 A portion of Poisson (1811, page 505, paragraph 3, lines 3, 9 and 11) showing the appearance of the word 'hélice' (French). (Reprinted with HathiTrust (https://babel.hathitrust.org/cgi/pt?id=hvd.hxj4gd&seq=539&view=1up. Accessed on the 27th of November 2025). Public domain)

Thomson (1856) stimulate any further works related to the mathematical studies of helicity?

10.3 The Mathematical Studies of Potential Helical Flows by Georg Karl Wilhelm Hamel in 1937

Georg Karl Wilhelm Hamel (1877–1954) was a German mathematician (the second person in Fig. 10.1). According to Wasserman (1958/1960, page 443, line 1), 'Potential helical flows have been completely described by G. Hamel [Hamel 1937].' Hamel's (1937) work has stimulated a number of studies of helical flows, including Wasserman's (1958/1960) work *Helical fluid flows*.

10.4 The Mathematical Study of Hydromagnetic Invariant Made by Walter Maurice Elsasser in 1956

As shown in Fig. 10.1 (the fourth person), Walter Maurice Elsasser (1904–1991) was a German-born American physicist who made such fundamental discoveries in physics, meteorology, and geophysics. In his review of *Hydromagnetic Dynamo Theory*, Elsasser (1956, page 150, equation (8.22)) obtains the following hydromagnetic invariant:

$$\int \mathbf{A}\cdot \boldsymbol{\upsilon} \times \mathbf{B} dV \tag{10.1}$$

where $\mathbf{A}$ is the magnetic vector potential; $\boldsymbol{\upsilon}$ the fluid velocity; $\mathbf{B}$ the field variable; and V the volume of a closed system.

Although the term is not explicitly used in Elsasser (1956, page 150), the above invariant is the so-called magnetic helicity H_m.

10.5 The Mathematical Studies of Magnetohydrodynamic Invariants by Lodewijk Woltjer in 1958

Lodewijk Woltjer (1930–2019) was a Dutch astronomer (the fifth person in Fig. 10.1). In his paper *A theorem on force-free magnetic fields*, Woltjer (1958a, page 489, equation (3)) shows that in a closed system

$$\int_V \boldsymbol{A}\cdot curl\, \boldsymbol{A} dV = \text{constant} \tag{10.2}$$

In his paper entitled *On hydromagnetic equilibrium*, Woltjer (1958b, page 833, equations (1–6)) presents the following integrals:

$$I_1 = \int_V \boldsymbol{A} \cdot curl\, \boldsymbol{A} dV \tag{10.3a}$$

$$I_2 = \int_V \boldsymbol{H} \cdot \upsilon dV \tag{10.3b}$$

$$I_3 = \int_V \boldsymbol{i} \times \boldsymbol{r} \cdot \rho v dV \tag{10.3c}$$

$$I_4 = \int_V \boldsymbol{j} \times \boldsymbol{r} \cdot \rho v dV \tag{10.3d}$$

$$I_5 = \int_V \boldsymbol{k} \times \boldsymbol{r} \cdot \rho v dV \tag{10.3e}$$

$$I_6 = \int_V \rho dV \tag{10.3f}$$

where ***i,j*** and ***k*** are unit vectors in arbitrary system of Cartesian co-ordinates, ***r*** the radius vector; and ***H*** the strength of the field (i.e. the magnetic field).

Although the terms are not explicitly used in Woltjer (1958b), the so-called magnetic helicity H_m is defined in Eqs. (10.2) and (10.3a) and the cross helicity H_c in Eq. (10.3b).

10.6 The Mathematical Studies of Helicity by Jean Jacques Moreau in 1959 and 1961

Jean Jacques Moreau (1923–2014) was a French mathematician, fluid dynamicist and engineer (the sixth person in Fig. 10.1). His short biography can be found in Alart et al. (2006, Preface, pages xiii–xiv).

Helicity is vorticity projected on the fluid velocity (Moreau 1959, page 772, lines 9 and 11 from the bottom) and is an 'invariant' (constant in time) of the Euler equations of fluid mechanics (cf energy) (Moreau 1961). It is a measure of the conserved 'degree of knottedness' of tangled vortex lines—*ergo* topological in character (Moffatt 1969).

The following lines can be found in Alart et al. (2006, Preface, page xiii/bottom line, page xiv/lines 1–2):

> The helicity invariant in the dynamics of ideal fluids, discovered by Jean Jacques Moreau in 1962 [1961], provides a starting point for the consideration of certain problems arising in fluid dynamics.

In his paper entitled *Invariants intégraux d'un ensemble de solutions des équations de l'hydrodynamique*, Moreau (1959, page 772, line 9 from the bottom) writes the following integral:

$$\mathrm{I} = \int_{D_t} \zeta_{ik} \mathrm{X}_i \wedge \mathrm{X}_k d\tau \tag{10.4}$$

where ζ_{ik} is vorticity.

In his paper *Constantes d'un îlot tourbillonnaire en fluide parfait barotrope*, Moreau (1961, page 2810) writes the following integral:

$$\mathbf{H} = \iiint_D \vec{u} \cdot \vec{\xi} d\tau \tag{10.5}$$

Although the term is not explicitly used in Moreau (1961, page 2810), the above invariant is the so-called kinetic helicity H_c.

What motivated Moreau to discover helicity? The notions "théorème de Kelvin", "équations de Helmholtz" and "tourbillonnaire (vortices in English)" appear a number of times in Moreau (1959, page 772; 1961, page 2811). Clearly, Kelvin's work on circulations and Helmholtz's equations of the vortex motion have motivated Moreau to discover helicity. The singular discovery by Jean-Jacques Moreau of the helicity invariant of the Euler equations has its roots in the celebrated laws of vortex motion (Moffatt 2018, page 165, line 1).

Does Moreau (1961) have any unnoticed points or implications? If we re-take a look at Moreau (1961), we may be able to see those notes in his paper. As shown in Fig. 10.3, Moreau (1961, page 2812, footnote (1), lines 6–9) seems to suggest the case of two linked vortex rings and anticipate (without proof) the correct value of helicity.

The lines 6–9 in Fig. 10.3 are translated into English by Renzo L. Ricca:

> …Evidently this is not the general case; as an example of a sub-region that has a non-zero H, we propose a system of two vortex rings, respectively of strength I_1 and I_2, linked, the whole embedded in an irrotational fluid to form a simply connected region: thus we obtain $H = 2I_1I_2$.

Despite Ricca's past limited advert, this note remains unnoticed. It would provide an updated perspective of the recent work on helicity and knot theory.

(¹) O. Bjorgum (*On Beltrami vector fields and flows*, Univ. Bergen Arbok. Naturvit. Rekke, 1951, nº 1), cité par C. Truesdell (*Math. Rev.*, 15, 1954, p. 569 et *The kinematics of vorticity*, Indiana University Press, Bloomington, Ind., U. S. A., 1954, p. 120) attire déjà l'attention sur cette intégrale, en remarquant que si $\vec{u}$ = grad h + f grad g (f, g, h : « potentiels de Monge », avec h univoque), H est essentiellement nul pour un îlot tourbillonnaire. Ce n'est évidemment pas le cas général; comme exemple d'îlot ayant un H non nul, nous proposons un système de deux anneaux tourbillonnaire de révolution, d'intensité respectives I_1 et I_2, *enlacés*, le tout plongé dans du fluide irrotationnel pour constituer un îlot D simplement connexe : on trouve alors H = 2 I_1 I_2.

Fig. 10.3 A portion of Moreau (1961, page 2812, footnote (1), lines 6–9) contains those suggesting the case of two linked vortex rings and anticipating (without proof) the correct value of helicity. (Reprinted from HalScience (https://hal.science/hal-01865239/file/Constantes_ilot_tourbillonnaire_Moreau_CRAS_1961.pdf. Accessed on the 27th of November 2025). Public domain)

10.7 The Mathematical Studies of Helicity by Robert Betchov in 1961

Robert Betchov (1919–1996) was a native of Switzerland (right in Fig. 10.1). His short biography can be found in an obituary (Szewczyk 1997, page 99). He was an eminent fluid mechanician. In his one-page short research note entitled *Semi-Isotropic Turbulence and Helicoidal Flows*, Betchov (1961, page 925, equation (6)) writes the following invariant giving a good measure of the helicity of the flow:

$$I = \epsilon_{i,j,k}\, u_i(x) \frac{\partial u_j(x)}{\partial x_k} = 6E(0) \tag{10.6a}$$

Betchov (1961, page 925, equations (9–11)) further defines the three helicoidal invariants:

$$I = \epsilon_{ijk}\, u_i \left(\partial u_j / \partial x_k\right) \tag{10.6b}$$

$$I_1 = \epsilon_{ijk}\, U_i \left(\partial U_j / \partial x_k\right) \tag{10.6c}$$

$$I_2 = \epsilon_{ijk}\, u_i' \left(\partial u_j' / \partial x_k\right) \tag{10.6d}$$

where the Levi-Civita symbol ϵ_{ijk} is the permutation tensor; u_i the Eulerian velocity; U_i the mean velocity; and u_i' the turbulent velocity.

10.8 The Mathematical Studies of Schraubensinn (German) Made by Max Christian Theodor Steenbeck and Fritz Krause in 1966

Max Christian Theodor Steenbeck (1904–1981) was a German astrophysicist (left in Fig. 10.4). Fritz Krause (1927–2024) was a German astrophysicist (center in Fig. 10.4).

Following up their previous study on the effects of a turbulent motion on magnetic fields, Steenbeck and Krause (1966) further give the consequences of this effect for rotating, electrically conducting spheres. For the first model, a sphere rotating with constant angular velocity, the quadratic effect provides for dynamo maintenance of the magnetic field of the earth. For the second model, differential rotation is included and this model may be applicable to magnetic stars. They also look for dynamo maintenance of alternating fields, e.g. the general magnetic field of the sun.

Fig. 10.4 Left: Max Christian Theodor Steenbeck (1904–1981). (Reprinted with from wikidata (https://www.wikidata.org/wiki/Q68510. Accessed on the 27th of November 2025). Public domain). Center: Fritz Krause (1927–2024). (Reprinted with permission from the Leibniz Institute for Astrophysics Potsdam (AIP) (https://www.aip.de. Accessed on the 27th of November 2025). Photo taken by AIP/R. Arlt). Right: H. Keith Moffatt. (Photo taken by the author)

According to Roberts (2021, page 221, lines 3 and 4), in their studies above, Steenbeck and Krause (1966, page 1296, left column, paragraph 3, line 1; right column, paragraph 1, line 2) use the following quantity to describe 'screwiness' or 'Schraubensinn (German)': $\boldsymbol{u} \cdot \nabla \times \boldsymbol{u}$.

Steenbeck and Krause (1966) was translated into English (Roberts and Stix 1971, pages 49–79). Interestingly, 'Schraubensinn (German)' in Steenbeck and Krause (1966) is directly translated into 'helicity' (Robert and Stix 1971, page 79, paragraph 2, line 1). Based on Roberts and Stix (1971), K.H. Radler also made his contribution to the turbulent dynamo.

In Ricca's view, Steenbeck, Krause and Radler contributed so much to the dynamo theory that it was almost inevitable they should run into the helicity concept during their early work.

10.9 The Mathematical Study of Helicity Made by H. Keith Moffatt in 1969

As already mentioned before, H. Keith Moffatt is a British applied mathematician (right in Fig. 10.4). He has made contributions to dynamo theory, vortex dynamics, turbulence, and all aspects of fluid dynamics. What motived his interest in helicity? What is his contribution?

Thomson (1856) proposes the helicoidal rotatory effect of transparent bodies on polarized light. Thomson (1869) suggested that vortex lines are frozen in any inviscid barotropic flow under conservative body forces, and that the topology of a vortex tube is conserved. Woltjer (1958a, page 489, equation (3); 1958b, page 833, equation (1)) obtains the invariance of a quantity, now known as '*magnetic helicity*', in a perfectly conducting fluid. Betchov (1961) recognizes the potential importance of helicity, but had no idea of its Euler-invariance. This invariance was later proved by Moreau (1961).

During his Part III course on Magnetodynamics of Fluids in 1967, it suddenly dawned on Professor Moffatt that Woltjer's result followed simply from the fact that the lines of force of the magnetic field are frozen in the fluid. Moffatt further realizes that there must be an analogous result for the nonlinear Euler equations reflecting the fact that vortex lines are similarly frozen-in in this context.

In Professor Moffatt's own words, it was Professor G.K. Batchelor (1920–2010) who recognizes the physical significance of Woltjer's second invariant (Woltjer 1958b, page 833, equation (2)) but Moffatt was not aware of the existence of Betchov (1961) and Moreau (1961) at that time. Note that Woltjer's second invariant is actually the cross helicity H_c.

Mainly following up Woltjer (1958a, b), Moffatt (1969, page 117, line 4) introduces an integral quantity, the helicity of the flow. He showed that it is an invariant of the Euler equations, precisely because of topological conservation. The integral is

$$I = \int \mathbf{u}(\mathbf{x}) \cdot \boldsymbol{\omega}(\mathbf{x}) dV \quad (10.7)$$

where $\mathbf{u}(\mathbf{x})$ is the velocity field, $\boldsymbol{\omega}(\mathbf{x})$ a vorticity distribution, and V the volume. From the integral (10.7) above, we can see that helicity (I) is closely associated with vorticity $\boldsymbol{\omega}(\mathbf{x})$.

10.10 Discussion

10.10.1 *How Were the Concepts of Vortex, Knot, and Helicity Embedded in Their Time?*

My soul is an entangled knot,
Upon a liquid vortex wrought
By Intellect, in the Unseen residing,
And thine doth like a convict sit,
With marlinspike untwisting it,
Only to find its knottiness abiding;
…….

By James Clerk Maxwell To Hermann Stoffkraft, Ph.D., the Hero of a recent work called "Paradoxical Philosophy." A Paradoxical Ode [After Shelley] in 1878 (Campbell and Garnett 1882, page 649/bottom, page 650/lines 1–4)

Do vortex, knot, and helicity have any connections with each other? If so, how? In what ways? From a historical perspective, those rhymes by Maxwell seem to give us the best answer to these questions. Moreover, according to Professor H. Keith Moffatt, FRSE, FRS, the invariance of *helicity* is embedded in this verse above. Perhaps Maxwell really was a 100 years ahead of his time! [Homage to James Clerk

Maxwell at https://www.clerkmaxwellfoundation.org/]. The notion 'knotted vortices' appears in Profile of Keith Moffatt (Ahmed 2014, page 3650, Section 2).

As presented in a footnote† in Moffatt (1969, page 118, bottom), his work on *The degree of knottedness of tangled vortex lines* has been stimulated by Gauss (1833), Crowell and Fox (1963), and Thomson (1869). From a historical perspective, conceptually and physically, the works by Woltjer (1958a, b), Moreau (1959, 1961), Betchov (1961), and Moffatt (1969) may be rooted in Thomson's (1856) work on the helicoidal effects of transparent bodies on polarized light.

Mathematically, the 'helicity' of the field, i.e. the mean asymptotic rotation of the phase curves around each other, is related to the asymptotic Hopf invariant (Arnold 1986, page 327, paragraph 2).

10.10.2 Schraubensinn (German) Versus Helicity

In their paper entitled *Homogeneous Dynamos: Theory and Practice*, Roberts and Jensen (1992) comment on Moffatt's (1969) work on helicity by adding the following lines after that reference:

> Here Moffatt played a role similar to that of George Johnson Stoney (who was responsible for christening J.J. Thompson's discovery the 'electron') by replacing the ponderous "Schraubensinn" of the Potsdam group by the felicious word "helicity". (Quoted from: https://aip.scitation.org/doi/10.1063/1.860703)

Notably, "the ponderous 'Schraubensinn' of the Potsdam group" refers to Steenbeck and Krause's (1966) work on the same mathematical expression of the same quantity but the different notions. 'Schraubensinn' is used by Steenbeck and Krause (1966) but 'helicity' by Moffatt (1969).

Subrahmanyan Chandrasekhar (1910–1995) was an Indian-born American astrophysicist who won the 1983 Nobel Prize for Physics. Woltjer published a joint paper with Chandrasekhar (Chandrasekhar and Woltjer 1958). We can cautiously infer that the mathematical study of magnetohydrodynamic invariants by Woltjer (1958a, b) might be influenced or inspired by Chandrasekhar. Chandrasekhar and Fermi (1953) was also cited in Steenbeck and Krause (1966). In his famous book entitled *Hydrodynamic and Hydromagnetic Stability* (Chandrasekhar 1961), 'The effect of fluid motions on the stability of helical magnetic fields' is presented in Section 114 in Chapter XII. Chandrasekhar's role in the mathematical studies of magnetohydrodynamic invariants and '*screwiness*' or '*Schraubensinn* (German)' cannot be underestimated.

10.10.3 *What Is the Interrelationship Between Chirality and Helicity?*

What is the interrelationship between chirality and helicity? What are their significances, e.g. in the theory of fluid dynamo? Moffatt (2018, page 168, Conclusion/3rd paragraph) gives the following answer:

> ...one of the great triumphs of the past half-century has been the discovery that chirality provides the key to understanding how this spontaneous growth can occur; and mean helicity provides the simplest indicator of this chirality

Helicity has its significance in helical magnetic field, solar cycles and solar dynamo (Zhang 2023, Chapters 5 and 6).

10.10.4 *What Is the Significance of Helicity in Turbulence?*

Does helicity have any other significances in the field of turbulence? The answer is yes. Below are a few examples:

(a) Brissaud et al. (1973) study helicity cascades in fully developed isotropic turbulence.
(b) The notion 'helical turbulence' also appears in the title (Kraichnan 1973).
(c) Helicity, as a physical quantity, has also been playing an important role in the establishment of turbulence closure models, in particular, modeling of turbulence in engineering (e.g. Liu et al. 2011, 2020). It has been used to account for the turbulence energy backscatter in the Spalart-Allmaras turbulence closure model in the region of corner separation in compressors (Liu et al. 2011). The helicity-corrected Spalart-Allmaras turbulence closure model has been implemented in various different types of engineering software, e.g. Turbostream (Kim et al. 2019), Hydra (Lopez et al. 2022), AU3D (Rao et al. 2022), and UPACS (Matsui et al. 2022).
(d) Similarly, helicity has also been used in shear stress transport turbulence models for predicting corner separation flow in linear compressor cascade (Liu et al. 2020).
(e) Independently, Biferale et al. (2012) also find the important role played by helicity in the energy transfer process.

10.11 Summary

- It is also interesting that there may be some significant works related to helicity before Thomson (1856), e.g., Poisson (1811).

- Simply because helicity is also one of the acts of nature among the most diverse phenomena, great physicists, like Maxwell, had an early perception of physical picture of helicity.
- The studies of magnetic and helicoidal effects of transparent bodies on polarized light by Lord Kelvin (Thomson 1856) might have also gradually stimulated the physical and mathematical studies of helicity by Elsasser, Woltjer, Moreau, and Moffatt.
- The mathematical studies of 'Schraubensinn (German)' may be independently made by Steenbeck and Krause.
- Amazingly, helicity has been important in all scientific and technological/engineering disciplines.

Postscript

Tsung-Dao Lee (1926–2024) was Nobel Prize laureate and Chinese-American physicist (the third person in Fig. 10.1). In his 1952 paper (Lee 1952), it is the first one that treats the Fourier coefficients of fluid and magnetofluid turbulence as a statistical canonical ensemble. In that paper, Lee defines the ideal probability density function (PDF). From a historical perspective, Lee (1952) knew of only one invariant of the energy, while the other three invariants in ideal hydrodynamics (the kinetic helicity H_k) and magnetohydrodynamics (the magnetic helicity H_M and the cross helicity H_c) were only discovered later. There is also another invariant, the parallel helicity H_P, which was not recognized until Shebalin (2006). From a historical perspective, Lee planted the initial seed. In this regard, for the details, please refer to Shebalin (2013, Section 2.1; 2024).

"*The Statistical Mechanics of Ideal Magnetohydrodynamic Turbulence and a Solution of the Dynamo Problem*", i.e. Shebalin (2024), is the most up to date work in this field. Elsässer (1956) is very important and was cited in Shebalin (2024, the third paragraph). In his paper, Shebalin uses what Elsässer wrote in his 1956 paper as a framework that he fills out. However, Shebalin (2024) is technical and perhaps difficult for the geodynamo community to absorb on only a cursory reading. Some problems remain, for example some consider the dynamo problem was solved by a technique called mean-field electrodynamics (MFE) 60 years ago while others consider it an open problem. Shebalin (2024) is important in presenting a new perspective on this debate and a potential solution.

However, since this chapter in this book mainly focuses on the early history of helicity, we may present in detail a critical review of these early mathematical studies of helicity above and the state-of-the art elsewhere.

Epilogue

> ...God is a mathematician of a very high order, and He used very advanced mathematics in constructing the universe. Dirac (1963, page 53, left column/paragraph 3, lines 15–18)

- Some elaborations are made on the historical development of the concepts of vortex, vorticity, knot, chirality, and helicity, with special reference to their early (or possibly original) conceptual, physical and mathematical ideas, and their mutual interrelationships.
- The diverse body of researchers contributing to the concepts and studies of vortex, vorticity, knot, chirality, and helicity include artists, natural philosophers, mathematicians, physicians, and physicists.
- All this early work is purely formal and somewhat mystifying. Originating from the differentials of the displacements in the analyses of fluid motion or light propagation, by a mathematical and physical analogy, the conception of the displacements has also been applicable to the theory of light, sound, heat, magnetism, electricity, and galvanic currents.
- Mathematical and physical analogies have played an important role in the studies of rotary displacements, velocities of rotation, or moments of couples producing rotation. In a way, they have stimulated each other.
- Lord Kelvin seems to have played more important roles than others in the physical and mathematical studies of vortex, vorticity, knot, chirality and helicity at the various stages.
- As a final note, the author cannot claim to be competent with all the quantities, aspects or topics presented in this book. However, throughout the writing of this book and the voyage from vortex to helicity, it is amazingly found that all the quantities, which are connected with each other in one way or another, have been gradually discovered and developed by those talented mathematicians and physicists in different eras, institutions and countries. The Creator is truly a mathematician of a very high order.

J. Z. Shi, *The Voyage from Vortex to Helicity (1517 – 1969)*, SpringerBriefs in History of Science and Technology, https://doi.org/10.1007/978-3-032-17853-4

References

Ackroyd P (2006) Isaac Newton. Chatto & Windus, London. 163 pp

Ahmed F (2014) Profile of Keith Moffatt. Proc Natl Acad Sci 111(10):3650–3652

Aiton EJ (1972) The vortex theory of planetary motions. Elsevier. ix + 282 pp

Alart P, Maisonneuve O, Rockafellar RT (2006) Nonsmooth mechanics and analysis. Springer. xiv + 320 pp

Aref H (2010) 150 years of vortex dynamics. Theor Comp Fluid Dyn 24:1–7

Arnold VI (1986) The asymptotic Hopf invariant and its applications. Sel Math Sov 5(4):327–345

Bernoulli D (1738) Hydrodynamica (in Latin). Typis Joh. Henr. Deckeri Typographi Bafilienfis. 304 pp + Tab. XII

Betchov R (1961) Semi-isotropic turbulence and helicoidal flows. Phys Fluids 4:925–926

Biferale L, Musacchio S, Toschi F (2012) Inverse energy cascade in three-dimensional isotropic turbulence. Phys Rev Lett 108:164501

Blackwell RJ (1966) Descartes' laws of motion. Isis 57(188):220–234

Blatter G, Feigel'man MV, Geshkenbein VB, Larkin AI, Vinokur VM (1994) Vortices in high-temperature superconductors. Rev Mod Phys 66:1125

Bottazzini U (1986) The higher calculus: a history of real and complex analysis from Euler to Weierstrass. Translated from the Italian by W. van Egmond. Springer. vii + 332 pp

Boussinesq J (1903) Théorie analytique de la Chaleur, vol II. Gauthier-Villars, Paris. XXXII and 625 pp

Brady R, Anderson R (2015) Maxwell's fluid model of magnetism. https://arxiv.org/pdf/1502.05926.pdf. Accessed 22 Sept 2025

Brewster D (1855a) Memoirs of the Life, Writings, and Discoveries of Sir Isaac Newton. Volume I. Edinburgh: Thomas Constable and Co., xv+487 pp

Brewster D (1855b) Memoirs of the Life, Writings, and Discoveries of Sir Isaac Newton. Volume II. Edinburgh: Thomas Constable and Co., xi+564 pp

Brissaud A, Frisch U, Leorat J, Lesieur M, Mazure A (1973) Helicity cascades in fully developed isotropic turbulence. Phys Fluids 16(8):1366–1377

Campbell L, Garnett W (1882) The Life of James Clerk Maxwell. With a selection from his correspondence and occasional writings and a sketch of his contributions to science. London, xvi+662 pp

Cauchy A-L (1815/1827) Théorie de la propagation des ondes à la surface d'un fluide pesant d'une profondeur indéfinie. Sciences mathématiques et physiques, vol I. Imprimé par autorisation du Roi à l'Imprimerie royale, pp 3–312

J. Z. Shi, *The Voyage from Vortex to Helicity (1517 – 1969)*, SpringerBriefs in History of Science and Technology, https://doi.org/10.1007/978-3-032-17853-4

Cauchy AL (1841) Mémoire sur les sommes alternées, connues sous le nom de résultantes. Exercises d'analyse et de physique mathématique 2, 160–176
Chandrasekhar S (1961) Hydrodynamic and hydromagnetic stability. Clarendon Press: OUP. 652 pp
Chandrasekhar S, Fermi E (1953) Problems of gravitational stability in the presence of a magnetic field. Astrophys J 118:116–141
Chandrasekhar S, Woltjer L (1958) On force-free magnetic fields. Proc Natl Acad Sci 44(4):285–289
Chorin AJ (1994) Vorticity and turbulence. Springer, New York. VIII + 176 pp
Clebsch RFA (1859) Ueber die Integration der hydrodynamischen Gleichungen. J Reine Angew Mathematik vi:1–10
Colagrossi A, Marrone S, Colagrossi P, Le Touze D (2021) Da Vinci's observation of turbulence: a French-Italian study aiming at numerically reproducing the physics behind one of his drawings, 500 years later. Phys Fluids 33:115122
Conway AW (1935/1936) Integrals of MacCullagh's equations. Proc R Irish Acad A43:25–33
Crowdy D, Tanveer S (2014) Philip Geoffrey Saffman. 19 March 1931–17 August 2008. Biogr Mem Fellows R Soc London 60:375–395
Crowell RH, Fox RH (1963) Introduction to knot theory. Springer, New York. x + 182 pp
d'Alembert JLR (1743) Traité de dynamique. Chez David, Paine, Libraire, rue & vis-à-vis la grille des Mathurins, A Paris. 272 pp + 5 plates
d'Alembert JLR (1744) Traité de l'équilibre et du mouvement des fluids. Chez David, Paine, Libraire, rue Saint Jacques, a la Plume d'or, A Paris. 458 pp + 10 plates
d'Alembert JLR (1749) Recherches sur la courbe que forme une corde tendüe mise en vibration. In: Histoire de l'Académie Royale des Sciences et des Belles-Lettres de Berlin. Année 1747. Haude et Spener, Berlin, pp 214–219
d'Alembert JLR (1752) Essai d'une nouvelle théorie de la résistance des fluides. Chez David l'aîné, Libraire, rue S. Jacques, la Plume d'or, Paris. Xlvj + 212 + 2 plates
Darrigol O (1998) From organ pipes to atmospheric motions: Helmholtz on fluid mechanics. Hist Stud Phys Biol Sci 29:1–51
Darrigol O (2005) Worlds of flow: a history of hydrodynamics from the Bernoullis to Prandtl. OUP. 356 pp
Darrigol O (2010) James MacCullagh's ether: an optical route to Maxwell's equation? Eur Phys J H35:133–172
Darrigol O, Frisch U (2008) From Newton's mechanics to Euler's equations. Phys D 237:1855–1869
Descartes R (1644) Principia Philosophiae. Amstelodami, Ludovicum Elzevirium, 310 pp
Des-Cartes (Descartes) R (1644) Principia philosophiae. Ludovicum Elzevirium, Amstelodami. 310 pp
Des-Cartes (Descartes) R (1664) Le Monde. Michel Bobin et Nicolas le Gras, Paris. 322 pp
Dirac PAM (1963) The evolution of the physicist's picture of nature. Sci Am 208(5):45–53
Eddington AS (1946) Fundamental theory. CUP. viii + 292
Elsasser WM (1956) Hydromagnetic dynamo theory. Rev Mod Phys 28(2):135–163
Euler L (1746) Nova theoria lucis et colorum. Opuscula Varii Argumenti 1:169–224
Euler L (1750/1752) Découverte d'un nouveau principe de Mécanique. Mém l'Acad Sci Berlin 6:185–217
Euler L (1752) Découverte d'un nouveau principe de Mécanique. Mémoires de l'Académie des Sciences de Berlin 6:185–217
Euler L (1755a) Principes généraux du mouvement des fluides. Mémoires de l'Académie des Sciences et Belles-Lettres de Berlin T11:217–273
Euler L (1755b) Principes généraux de l'etat d'equilibre des fluides. Mémoires de l'Académie des Sciences et Belles-Lettres de Berlin T11:274–315
Euler L (2008) General principles of the motion of fluids. Phys D 237:1825–1839
Falconer I (2019) Vortices and atoms in the Maxwellian era. Philos Trans R Soc Lond A377:20180451
Faraday M (1846) Thoughts on ray-vibrations. Philos Mag 3 28(188):345–350
Filon LNG (1929) Obituary Notice. Micaiah John Muller Hill. Proc R Soc Lond A124(795):i–v

Florides PS (2008) John Lighton Synge. 23 March 1897–30 March 1995. Biogrl Mem Fellows R Soc London 54:401–424
Fourier JBJ (1822) Théorie analytique de La Chaleur (in French). Chez Firmin Didot, Pere et Fils, Paris, p 639
Fourier JBJ (1878) The analytical theory of heat. Translated, with Notes, by Alexander Freeman. CUP. 466 pp
Fraser C (1985) Lagrange's changing approach to the foundations of the calculus of variations. Arch Hist Exact Sci 32(2):151–191
Frisch U, Villone B (2014) Cauchy's almost forgotten Lagrangian formulation of the Euler equation for 3D incompressible flow. arXiv:1402.4957 Accessed 23 Sept 2025
Gauss CF (1833) Zur mathematischen Theorie der electrodynamischen Wirkungen. Werke. Koniglichen Gesellschaft der Wissenschaften zu Gottingen 1877(5):605
Hamel G (1937) Potentialstromungen mit konstanter Geschwindigkeit. Sitzungsberichte der Preussischen Akademie der Wissenschaften zu Berlin, Physikalisch-Mathematische Klasse, pp 5–20
Hamilton WR (1824/1828) Theory of systems of rays. Trans R Irish Acad 15:69–174
Hamilton WR (1831/1837) On differences and differentials of functions of zero. Trans R Irish Acad 17:235–236
Hamilton WR (1832) Third supplement to an essay on the theory of systems of rays. Trans R Irish Acad 17:v–ix + 144
Hamilton WR (1840/1843) On fluctuating functions. Trans R Irish Acad 19:264–321
Hamilton WR (1847) On Quaternions. Proceedings of the Royal Irish Academy III:273–292
Hardy GH (2008) A course of pure mathematics, 10th edn/Centenary edn. CUP. 530 pp
Heimann PM (1972) The unseen universe: physics and the philosophy of nature in Victorian Britain. Br J Hist Sci 6:73–79
Helmholtz H (1858) Über Integrale der hydrodynamischen Gleichungen, welche den Wirbelbewegungen entsprechen (in German). J Mathematik 1:25–55
Helmholtz H (1867) On integrals of the hydrodynamic equations, which express vortex-motion. [Translated by P. G. Tait]. Philos Mag Ser 4 33(226):485–512
Helmholtz H (1978) On integrals of the hydrodynamic equations that correspond to vortex motions. [Translated by Uwe Parpart]. Int J Fusion Energ 1(3–4):41–68
Herschel JFW (1815/1816) On the development of exponential functions; together with several new theorems relating to finite differences. Philos Trans R Soc Lond 106:25–45
Hicks WM (1899) Researches in vortex motion. – Part III. On spiral or gyrostatic vortex aggregates. Philosophical Transactions of the Royal Society of London 192:33–99
Hill MJM (1884) On the motion of fluid, part of which moving rotationally and part irrotationally. Philos Trans R Soc Lond 175:363–409
Hill MJM (1894) On a spherical vortex. Philosophical Transactions of the Royal Society of London A185:213–245
Jellett JH, Haughton S (1880) The collected works of James MacCullagh, LLD. Dublin University Press Series, Dublin. v + 381 pp
Jorink E, Maas A (2012) Newton and the Netherlands. How Isaac Newton was fashioned in the Dutch Republic. Leiden University Press. 256 pp
Kauffmann LH (2000) Knots and physics. World Scientific, Singapore. 788 pp
Keynes M (1995) The personality of Isaac Newton. Notes and Records of the Royal Society 49(1):1–56
Kim S, Pullan G, Hall CA, Grewe RP, Wilson MJ, Gunn E (2019) Stall inception in low-pressure ratio fans. J Turbomach 141:071005–071001
Knott CG (1911) Life and scientific work of Peter Guthrie Tait. Cambridge University Press. x + 379 pp
Koenigsberger L (1906) Hermann von Helmholtz. Translated by Frances A. Welby. Oxford at the Clarendon Press. xvii + 440 pp
Kragh H (2002) The vortex atom: a Victorian theory of everything. Centaurus 44:32–114

Kraichnan RH (1973) Helical turbulence and absolute equilibrium. J Fluid Mech 59(4):745–752
Lagrange J-L (1760/1761) Application de la méthode exposée dans le Mémoire précédent à la solution de différents problèmes de dynamique t. II:365–468
Lagrange J-L (1781) Mémoire sur la théorie du mouvement des fluides. Nouveaux mémoires de l'Académie royale des sciences et belles-lettres de Berlin, année 1781. Œuvres Complètes 4:695–748
Lagrange J-L (1788) Mécanique analytique. Chez La Veuve Desaint, Libraire, rue du Foin S. Jacques, A Paris. xii + 512 pp
Lagrange J-L (1815) Mécanique analytique, vol II. MmeVe Courcier, Imprimeur-Libraire Pour Les Mathematiques, Paris. viii + 378 pp
Lamb H (1879) A Treatise on the mathematical theory of the motion of fluids, 1st edn. CUP. 258 pp
Lamb H (1895) Hydrodynamics, 2nd edn. CUP. vii + 604 pp
Lamb H (1906) Hydrodynamics, 3rd edn. CUP. xv + 634 pp
Lamb H (1916) Hydrodynamics, 4th edn. CUP. xvi + 708 pp
Lamb H (1924) Hydrodynamics, 5th edn. CUP. xvi + 687 pp
Lamb H (1932) Hydrodynamics, 6th edn. CUP. xv + 738 pp
Launder BE (2012) Horace Lamb and the circumstances of his appointment at Owens College. Notes Rec Roy Soc 67:139–158
Lee TD (1952) On some statistical properties of hydrodynamical and magneto-hydrodynamical fields. Q Appl Math 10:69–74
Liebling TM, Pourin L (2012) Voronoi diagrams and Delaunay triangulations: ubiquitous Siamese twins. Documenta Mathematica-Extra ISMP:419–431
Liu YW, Lu LP, Fang L, Gao F (2011) Modification of Spalart-Allmaras model with consideration of turbulence energy backscatter using velocity helicity. Phys Lett A 375:2377–2381
Liu YW, Tang YM, Scillitoe AD, Tucker PG (2020) Modification of shear stress transport turbulence model using helicity for predicting corner separation flow in linear compressor cascade. J Turbomach 142:021004–021001
Lopez DI, Ghisu T, Kipouros T, Shahpar S, Wilson M (2022) Extending highly loaded axial fan operability range through novel blade design. J Turbomach 144:121009–121001
Lord Kelvin (1884/1904) Baltimore lectures on molecular dynamics and the wave theory of light. Cambridge University Press, Clay and Sons, xxi + 703 pp
Love AEH, Glazebrook RT (1935) Sir Horace Lamb, 1849–1934. Obit Not Fell R Soc 1:374–392
Lugt HJ (1979) The dilemma of defining a vortex. In: U Muller, KG Roesner and B Schmidt (eds.), Recent Developments in Theoretical and Experimental Fluid Mechanics Compressible and Incompressible Flow., Springer-Verlag Berlin Heidelberg New York, pp. 309–321
Lugt HJ (1983) Vortex flow in nature and technology. Wiley, New York. 297 pp
Lugt HJ (1985) Vortices and vorticity in fluid dynamics: the development of vortices may serve as a paradigm to illustrate how patterns in nature become organized. Am Sci 73:162–167
Lugt HJ (1996) Introduction to vortex theory. Vortex Flow Press, Potomac, MD. 627 pp
M'Kendrick JG (1899) Herman Ludwig Ferdinand von Helmholtz. T. Fisher Unwin, London. xvi + 299 pp
MacCullagh J (1839/1846) An essay towards a dynamical theory of crystalline reflexion and refraction. Trans R Irish Acad 21:17–50
MacCullagh J (1841) On the dynamical theory of crystalline reflexion and refraction. Proc R Irish Acad 2:96–103
MacCullagh J (1842) On the dispersion of the optic axes, and of the axes of elasticity, in biaxel crystals. Philos Mag 3(21):293–297
Marusic I, Broomhall S (2021) Leonardo da Vinci and fluid mechanics. Annu Rev Fluid Mech 53:1–25
Matsui K, Tani N, Perez E, Kelly RT, Jemcov A (2022) Calibrated rotation-helicity-quadratic constitutive relation Spalart-Allmaras (R-H-QCR SA) model for the prediction of multi-stage compressor characteristics. In: Proceedings of ASME turbo expo 2022 turbomachinery technical conference and exposition GT2022, June 13–17, 2022, Rotterdam, The Netherlands, pp 1–9

Maucher F, Gardiner SA, Hughes IG (2016) Excitation of knotted vortex lines in matter waves. New J Phys 18:063016
Maxwell JC (1861a) On physical lines of force. Part I. Philos Mag 21(141):338–348
Maxwell JC (1861b) On physical lines of force. Part II. Philos Mag 21(139):161–175
Maxwell JC (1862) On physical lines of force. Part IV. Philos Mag 23(152):85–95
Maxwell JC (1873a) A treatise on electricity and magnetism, vol I. Clarendon Press, Oxford. xxix + 425, Plates I–XIII
Maxwell JC (1873b) A treatise on electricity and magnetism, vol II. Clarendon Press, Oxford. xxiii + 444, 7 Plates
Meleshko VV (2010) Coaxial axisymmetric vortex rings: 150 years after Helmholtz. Theor Comput Fluid Dyn 24:403–431
Meleshko VV, Aref H (2007) A bibliography of vortex dynamics 1858–1956. Adv Appl Mech 41:197–292
Meleshko VV, Gourjii AA, Krasnopolskaya TS (2012) Vortex rings: history and state of the art. J Math Sci 187(6):772–808
Milne EA (1947) Book review of fundamental theory. Nature 159:486–488
Milner SR (1934) William Mitchinson Hicks (1850–1934). Obit. Not. R Soc. Lond. 1:393–339
Moffatt HK (1969) The degree of knottedness of tangled vortex lines. J Fluid Mech 35:117–129
Moffatt K (2008) Vortex dynamics: the legacy of Helmholtz and Kelvin. In: Borisov AV (ed) IUTAM symposium on Hamilton dynamics, vortex structures, turbulence. Springer, pp 1–10
Moffatt HK (2010) A brief introduction to vortex dynamics and turbulence. In: Moffatt HK, Shuckburgh (eds) Environmental hazards the fluid dynamics and geophysics of extreme events. World Scientific, pp 1–27
Moffatt HK (2017) Corrigendum. The degree of knottedness of tangled vortex lines. J Fluid Mech 830:821–822
Moffatt HK (2018) Helicity. CR-Mec 346(3):165–169
Moffatt HK, Dormy E (2019) Self-exciting fluid dynamos. Cambridge University Press. xviii + 520 pp
Moffatt HK, Tsinober A (1990) Topological fluid mechanics. In: Proceedings of the IUTAM symposium, Cambridge, UK, 13–18 August 1989. Cambridge University Press. 823 pp
Moreau JJ (1959) Invariants intégraux d'un ensemble de solutions des équations de l'hydrodynamique (in French). Comptes rendus de l'Académie des sciences Paris 248:771–773
Moreau JJ (1961) Constantes d'un îlot tourbillonnaire en fluide parfait barotrope. C R Acad Sci Paris 252:2810–2812
Newton I (1687) Philosophiae naturalis principia mathematica (in Latin). Royal Society of London. 510 pp
Newton I (1846) The mathematical principles of natural philosophy. Daniel Adee, New York. lxix + 72–581 pp
Newton I (1999) The principia: mathematical principles of natural philosophy (1726), 3rd edn. University of California Press, Berkeley. 1025 pp
Nitsche M (2006) Vortex dynamics. In: Francoise J-P, Naber GL, Tsun TS (eds) Encyclopedia of mathematics and physics. Elsevier Science Ltd., pp 390–399
Noll W (2003) The genesis of Truesdell's nonlinear field theories of mechanics. J Elast 70:23–30
Palle D (2021) On chirality of the vorticity of the Universe. https://arxiv.org/pdf/0802.2060.pdf
Poisson S-D (1811) Traité de mécanique, vol 1. Vve Courcier, Paris. xviii + 507 + 5 plates
Polya G (1954) Mathematics and plausible reasoning, vol I. Princeton University Press. xvi + 280 pp
Power I (1842/1843) On the truth of the hydrodynamical theorem. Trans Cambridge Philos Soc VII (III):455–463
Pullin DI, Saffman PG (1998) Vortex dynamics in turbulence. Annu Rev Fluid Mech 30:31–51
Ranford P (2020) Sir George Gabriel Stokes, Bart (1819–1903): his impact on science and scientists. Philos Trans R Soc Lond A378:20190524

Rankine WJM (1851) On the centrifugal theory of elasticity, as applied to gases and vapours. Philos Mag 4 2(14):509–542
Rankine WJM (1855a) On the hypothesis of molecular vortices, or centrifugal theory of elasticity, and its connexion with the theory of heat. Philos Mag 4 X(67):354–363
Rankine WJM (1855b) On the hypothesis of molecular vortices, or centrifugal theory of elasticity, and its connexion with the theory of heat. Philos Mag X(68):411–420
Rankine WJM (1870) On the thermal energy of molecular vortices. Philos Mag 39(260):211–221
Rao AN, Sureshkumar P, Stapelfeldt S, Lad B, Lee K-B, Rico RP (2022) Unsteady analysis of aeroengine intake distortion mechanisms: vortex dynamics in crosswind conditions. In: Proceedings of ASME turbo expo 2022 turbomachinery technical conference and exposition GT2022, June 13–17, 2022, Rotterdam, The Netherlands, GT2022-81805
Roberts P (2021) Review of self-exciting fluid dynamos, by H. Keith Moffatt and Emmanuel Dormy, Cambridge University Press, 2019, 520 pp. Geophys Astrophys Fluid Dyn 115(2):221–230
Roberts PH, Jensen TH (1992) Homogeneous dynamos: theory and practice. Phys Fluids B5:2657–2662
Roberts PH, Stix M (1971) The turbulent dynamo. A translation of a series of papers by F. Krause, K.-H. Radler, and M. Steenbeck. National Center for Atmospheric Research-TN/IA-60. 318 pp
Robinson JM (1968) An introduction to early Greek philosophy: the chief fragments and ancient testimony, with connecting commentary. Houghton Mifflin, Boston. x + 342 pp
Roget PM (1847) Obituary of James MacCullagh. Proc R Soc Lond 69:712–718
Rowland HA (1880) On the motion of a perfect incompressible fluid when no solid bodies are present. Am J Math 3(3):226–268
Saffman PG (1981) Dynamics of vorticity. J Fluid Mech 106:49–58
Saffman PG (1992) Vortex dynamics. CUP. 311 pp
Saffman PG (1997) Vortex models of isotropic turbulence. Philos Trans R Soc Lond A355:1949–1956
Saffman PG, Baker GR (1979) Vortex interaction. Annu Rev Fluid Mech 11:95–122
Sandifer (ed) (2010) How Euler did it. http://eulerarchive.maa.org/hedi/HEDI-2010-02.pdf. Accessed 23 Sept 2025
Scaife KP (1990) James MacCullagh, M.R.I.A., F.R.S., 1809–47. Proc R Irish Acad Hist Literat 90C:67–106
Serrin J (1959) Mathematical principles of classical fluid mechanics. In: Truesdell C (ed) Encyclopedia of physics, PHYSIK 3, Fluid dynamics, vol I. Springer-Verlag OHG, pp 125–263
Shebalin JV (2006) Ideal homogeneous magnetohydrodynamic turbulence in the presence of rotation and a mean magnetic field. Aust J Plant Physiol 72:507–524
Shebalin JV (2013) Broken ergodicity in magnetohydrodynamic turbulence. Geophys Astrophys Fluid Dyn 107(4):411–466
Shebalin JV (2024) The statistical mechanics of ideal magnetohydrodynamic turbulence and a solution of the dynamo problem. Fluids 9:46
Shi JZ (2021a) George Keith Batchelor's interaction with Chinese fluid dynamicists and inspirational influence: a historical perspective. Notes Rec R Soc 75:461–502
Shi JZ (2021b) Qian Jian (1939–2018) and his contribution to small-scale turbulence studies. Phys Fluids 33(4):041301
Shi JZ (2024a) Revd. Dr. John C. Polkinghorne's activities in science and religion. Eur J Sci Theol 20(3):1–33
Shi JZ (2024b) Some early studies of isotropic turbulence: a review. Atmosphere 15(4):494
Shi JZ (2024c) John C. Polkinghorne and Chen Ning Yang on the dialogue between science and religion. Christian Perspect Sci Technol New Ser 3:209–257
Silver DS (2014) Knots in the nursery: (cats) cradle song of James Clerk Maxwell. Not AMS 61(10):1186–1194
Soni V, Billign ES, Magkiriadou S, Sacanna S, Bartolo D, Shelley MJ, Irnine WTM (2019) The odd free surface flows of a colloidal chiral fluid. Nat Phys 15:1188–1194
Spearman TD (1995) William Rowan Hamilton, 1805–1865. Proc R Irish Acad 95A:1–12
Spearman TD (2010) James MacCullagh 1809–1847. Eur Phys J H 35:113–122

Steenbeck M, Krause F (1966) Erklärung stellarer und planetarer magnetfelder durch einen turbulenzbedingten dynamomechanismus. Z Naturforschung 21a:1285–1296

Stewart B, Tait PG (1876) The unseen universe or physical speculations on a future state. MacMillan and Co., New York. xii + 212 pp

Stokes GG (1842/1843) On the steady motion of incompressible fluids. Trans Cambridge Philos Soc VII(I):439–453

Stokes GG (1843) On some cases of fluid motion. Trans Cambridge Philos Soc VIII(I):105–137

Stokes GG (1845) On the theories of the internal friction of fluids in motion, and of the equilibrium and motion of elastic solids. Trans Cambridge Philos Soc VIII 22:287–342

Synge JL, Lin CC (1943) On a statistical model of isotropic turbulence. Trans R Soc Canada 37:45–79

Szewczyk AA (1997) Robert Betchov. Phys Today 50(3):99

Tait PG (1877) On knots. Part I. Trans R Soc Edinb XXVIII:145–190 + Plate XVI

Tait PG (1884) On knots. Part II. Trans R Soc Edinb XXXII:327–342 + Plate XLIV

Tait PG (1885) On knots. Part III. Trans R Soc Edinb XXXII:493–506 + Plates LXXIX, LXXX, and LXXXI

Taylor GI (1932) The transport of vorticity and heat through fluids in turbulent motion. Proc Roy Soc London A135:685–705

Taylor GI (1970) Some early ideas about turbulence. J Fluid Mech 41(1):3–11

Taylor GI, Green AE (1937) Mechanism of the production of small eddies from large ones. Proc Roy Soc London A158:499–521

Thomson W (1847) On a mechanical representation of electric, magnetic, and galvanic forces. Camb Dublin Math J II:61–64

Thomson W (1856) Dynamical illustrations of the magnetic and the helicoidal rotatory effects of transparent bodies on polarized light. Proc Roy Soc Lond 8:150–158

Thomson W (1867a) On vortex atoms. Proc R Soc Edinb, Session 1866–67 vi:94–105

Thomson W (1867b) The translatory velocity of a circular vortex ring. Philos Mag 4 XXIII(226):511–512

Thomson W (1869) On vortex motion. Trans R Soc Edinb XXV(1):217–260

Thomson W (1871) On the motion of free solids through a liquid. Proc R Soc Edinb 7:384–390

Thomson W (1875/1878) Vortex statics. Proceedings of the Royal Society of Edinburgh IX (94):59–73. [reprinted in The London, Edinburgh, and Dublin Philosophical Magazine and Journal of Science, 10:60, 97–109.]

Thomson W (1887a) On the stability of steady and of periodic fluid motion. The London, Edinburgh, and Dublin Philosophical Magazine and Journal of Science 5, XXIII (144):459-464

Thomson JJ (1883) A treatise on the motion of vortex rings. Cambridge University Press. xx + 124 pp

Truesdell C (1954) The kinematics of vorticity. Indiana University Press, Bloomington. xvii + 232 pp

van Jaarsveld JPJ, Holten APC, Elsenaar A, Trieling RR, van Heijst GJF (2011) An experimental study of the effect of external turbulence on the decay of a single vortex and a vortex pair. J Fluid Mech 670:214–239

Wasserman RH (1958/1960) Helical fluid flows. Q Appl Math XVII(4):443–445

Whittaker E (1954) William Rowan Hamilton. Sci Am 190(5):82–87

Woltjer L (1958a) A theorem on force-free magnetic fields. Proc Natl Acad Sci 44(6):489–491

Woltjer L (1958b) On hydromagnetic equilibrium. Proc Natl Acad Sci 44(9):833–841

Wu J-Z, Ma H-Y, Zhou M-D (2006) Vorticity and vortex dynamics. Springer, Berlin. XIV + 776 pp

Yuan J (2019) Vorticity induced by chiral plasmonic fields. Nat Mater 18:533–535

Zhang H (2023) Solar magnetism. Springer. xvii + 408 pp

Index

J. Z. Shi, *The Voyage from Vortex to Helicity (1517 – 1969)*, SpringerBriefs in History of Science and Technology, https://doi.org/10.1007/978-3-032-17853-4

The manufacturer's authorised representative in the EU is Springer Nature Customer Service Centre GmbH, Europaplatz 3, 69115 Heidelberg, Germany. If you have any concerns regarding our products, please contact ProductSafety@springernature.com

Printed and bound by CPI Group (UK) Ltd, Croydon, CR0 4YY
07/07/2026
02160927-0005